The Essential Guide to Passing the Geotechnical Civil PE Exam Written in the form of Questions

160 CBT Questions Every PE Candidate Must Answer

First Edition

Jacob Petro
PhD, PMP, CEng, PE

Request latest Errata, or add yourself to our list for future information about this book by sending an email with the book title, or its ISBN, in the subject line to:
Errata@PEessentialguides.com
or
Info@PEessentialguides.com

The Essential Guide to Passing the Geotechnical Civil PE Exam Written in the form of Questions

160 CBT Questions Every PE Candidate Must Answer

First Edition

Jacob Petro
PhD, PMP, CEng, PE

PE Essential Guides
Hillsboro Beach, Florida

> # Report Errors For this Book
>
> We are grateful to every reader who notifies us of possible errors. Your feedback allows us to improve the quality and accuracy of our products.
>
> Report errata by sending an email to Errata@PEessentialguides.com

The Essential Guide to Passing the Geotechnical Civil PE Exam Written in the form of Questions – 160 CBT Questions Every PE Candidate Must Answer

First Edition Print 1.1

© 2024 Petro Publications LLC. All rights reserved.

All content is copyrighted by Petro Publications LLC and its owners. No part, either text or image, may be used for any purpose other than personal use. Reproduction, modification, storage in retrieval system or retransmission, in any form or by any means, electronic, mechanical, or otherwise, for reasons other than personal use, without prior permission from the publisher is strictly prohibited.

For written permissions contact: Permissions@PEessentialguides.com

For general inquiries contact: Info@PEessentialguides.com

Imprint name: PE Essential Guides

Company owning this imprint: Petro Publications LLC. Established in Florida, 2023.

ISBN: 979-8-9891857-9-5

Release History

Date	Edition No.	Description and Update
September 2024	1	

Disclaimer

The information provided in this book is intended solely for educational and illustrative purposes. It is important to note that the technical information, examples, and illustrations presented in this book should not be directly copied or replicated in real engineering reports or any official documentation.

While there may be resemblances between the examples in this book and real structures, users must exercise caution and conduct comprehensive verification of all information before implementing it in any practical setting. The author and all affiliated parties explicitly disclaim any responsibility or liability arising from the misuse, misinterpretation, or misapplication of the information contained in this book.

Furthermore, it is essential to understand that this book does not constitute legal advice, nor can it be considered as evidence or exhibit in any court of law. It is not intended to replace professional judgment, and readers are encouraged to consult qualified experts or seek legal counsel for any specific legal or technical matters.

By accessing and utilizing the information in this book, readers acknowledge that they do so at their own risk and agree to hold the author and all affiliated parties harmless from any claims, damages, or losses resulting from the use or reliance upon the information provided herein.

Important Information About Printing and Quality Assurance

Most of our publications are printed through third-party services. While we have full trust in our partners, we acknowledge that occasional minor issues, such as missing pages or other errors, may arise. If you encounter any such issues with your copy, please do not hesitate to contact us via any of the emails listed on the copyright page, and we will be happy to assist you in resolving the matter.

Preface

General Information about the Book

This book is designed to help civil engineers pass the NCEES exam with its 2024 updated specifications, which is a prerequisite for obtaining the professional engineering PE license in the United States 2024 onwards. This book is tailored to provide you with comprehensive knowledge, detailed examples, and step-by-step solutions with ample graphics that are directly related to the subjects covered by the NCEES exam.

In this book, you will find an extensive collection of civil engineering problems that are carefully selected to build your knowledge, skills, and ability to apply fundamental principles and advanced concepts in the field of civil engineering. These problems are accompanied by detailed explanations, diagrams, and equations to help you understand the underlying principles and solve the problems efficiently and accurately.

Whether you are a recent graduate, an experienced engineer, or a professional who wants to obtain the engineering license in the United States, this book will prepare you for the exam and equip you with the necessary tools to succeed.

The book is structured in a way such that it provides the reader with a comprehensive understanding of the core topics that are covered by the NCEES exam 2024 specifications – which have more in-depth focus compared to the previous versions of the same exam.

The book provides the reader with a full coverage and understanding for the NCEES relevant exam topics and possible question scenarios during a real test situation. If certain topics or methods were not covered in this book, the book method of presentation will ultimately guide you on how to find the solution you seek on your own, know where to find it, and provide the solution on a timely manner that saves you time during the real exam.

The questions in this book are neither easy nor difficult. They are constructive and creative in nature. They have been authored in a way to have you remember the core concept of the engineering topic you seek. They are designed to challenge engineers to think critically and apply their knowledge to exam and real-world scenarios. These questions require a deeper level of analysis and understanding than simple recall of information. They may involve multiple steps or require the engineer to consider different perspectives or solutions, while they can be difficult and require significant amount of effort and research to solve them.

Reasons I wrote this Book

I decided to author this book because I have a strong passion for engineering. I have a deep interest and understanding of civil engineering, and I wanted to share this knowledge and passion with others.

I am also very enthusiastic about engineering, and this has allowed me to explore various concepts and develop unique perspectives.

By writing this book I hope I can inspire others to pursue and improve on their career in engineering, help them pass the NCEES exam, improve on their skills and advance their knowledge in this field and provide them with the tools they need to succeed.

Lastly, the energy and enthusiasm I have, and I brought into this work is infectious and I wanted to channel this energy into this fascinating project and share it with others. I strongly believe this book will be a valuable resource for anyone interested in learning more about civil engineering.

Acknowledgement and Dedication

I would like to thank the readers of this book, who I hope will find it informative, engaging, and thought-provoking. It is my sincere hope that this book will inspire others to pursue their own passion, and that it will serve as a valuable resource for all those interested in the field of engineering.

Table of Contents

Preface vii

    General Information about the Book vii
    Reasons I wrote this Book vii
    Acknowledgement and Dedication viii

Introduction 1

    General Description of this Book 1
    Book Structure 1
    Case Studies 2
    Theory Explanation 2

Book Parts 2

About the Exam 3

    General information 3
    Dissecting the Exam 4

How to use this Book 4

    General 4
    Which References to Own 5
    References and Bibliography 5

2024 Exam Specifications 5

Map of Problems Presented 8

Problems & Solutions 13

Site Charecterization 15
Soil Mechanics 33
Construction Observation 53
Earthquake Engineering 75
Earth Structures 89
Groundwater & Seepage 113
Problematic Soil 125
Walls & Foundations 137

References & Bibliography 181

References 183
Bibliography 185
A. Design of Earth Structures (Piles, Anchors, Sheet Piles, etc) 185
B. Designing with Geosynthetics 185
C. Earthquake Engineering 186
D. General Material 186
E. Ground Improvement 186

About the Author

Dr. Petro is a professional engineer and a business leader with over 20 years of experience in leading and growing engineering companies. Throughout his career, he worked with some of the most prestigious engineering firms. With a vast background in design and construction, he earned a reputation for delivering innovative and cutting-edge projects throughout his career. Dr. Petro is also the author of the series of the PE Essential Guides, which is a best seller and has comprehensive coverage for all civil engineering disciplines.

Dr. Petro is a civil engineer, he holds a Doctorate degree, he has earned a Professional Engineering (PE) license as well as Chartered Engineer (CEng) certification. Additionally, Dr. Petro has earned a Project Management Professional (PMP) certification, further demonstrating his expertise in managing complex projects. Over the years, he successfully led and managed teams of engineers, designers, and other professionals, overseeing complex projects from conception to completion.

Throughout his career, Dr. Petro designed and delivered numerous innovative and interesting projects that have contributed significantly to various industries he has worked in. His passion for engineering and business has driven him to publish several papers and articles in industry-leading journals and magazines. His work has been recognized as state-of-the-art and has been referenced by many industry professionals.

As an international civil engineer who worked across the globe, Dr. Petro brings an interesting perspective to the table. He has a deep understanding of how civil facilities and structures work and how to optimize them for maximum efficiency and safety. His ability to communicate complex engineering concepts to both technical and non-technical stakeholders has been key to his success.

This page is intentionally left blank

Introduction

General Description of this Book

The Essential Guide to Passing The Geotechnical Civil PE Exam is a guide designed in the form of questions. It aims to achieve a comprehensive coverage for the geotechnical discipline, currently being the most diverse while holding the lowest passing rates in the NCEES record – 46% passing rates for first half of 2023, 57% for the first half of 2024 and 38% for retakers.

This book offers a new way of delivery, it attempts to cover numerous exam scenarios with the use of questions. It aims to achieve good coverage for the required material in the geotechnical engineering discipline with the use of detailed and well referenced solutions which explain theory, offers case studies along with alternative solutions as it delivers the required concepts and material.

Given the diverse nature of geotechnical engineering, it is unlikely that all relevant information can be found in a single source. This have been done for you. The necessary research has been carried out and important information from various resources has been gathered from multiple sources to provide you with a good coverage for the NCEES required knowledge areas creating a one stop shop for your studies.

Although the guide comprises 160 questions, some require dual outcomes for a solution, effectively making them worth two or more questions. This intentional design ensures that interconnected concepts are delivered together, supporting the book comprehensive coverage strategy.

Additionally, you will find numerous conceptual questions and case studies designed to mimic exam scenarios and enhance understanding of key geotechnical concepts. The book also includes plenty of numerical questions.

The answers to the questions are detailed and they are well referenced. Some answers provide several methodologies in their presentation for the readers' benefit. Questions and answers are accompanied by graphics and detailed step-by-step explanations to help engineers understand concepts better. Not only this, the aim of providing such detailed answers is to help engineers understand all methods and apply them during the exam so they are better equipped with all possible exam scenarios.

Book Structure

The 2024 geotechnical civil PE exam specifications consist of ten knowledge areas. Those knowledge areas are presented in the relevant section of this book.

For ease of reference, and in order to ensure a clear, and a complete coverage for all possible question scenarios, those knowledge areas are grouped into eight parts in this book as follows:

Book Part		No. of Questions
I	Site Characterization	20
II	Soil Mechanics	20
III	Construction Observation	25
IV	Earthquake Engineering	15
V	Earth Structures	20
VI	Groundwater & Seepage	10
VII	Problematic Soil	15
VIII	Walls & Foundations	35

The number of questions in each part was carefully determined in a way that speaks to the expected number of questions in a real exam, also, they have been made as such ensuring good coverage for the intended material behind relevant knowledge areas for each part.

Case Studies
Geotechnical engineering is a diverse field that encompasses both innovative and traditional solutions. This book includes over 20 interesting and beneficial case studies covering various topics such as underpinning, ground grouting, slope stabilization, and the design of geosynthetics. Readers are encouraged to check the bibliography section for additional references that may provide further valuable case studies. Additionally, conducting your own research for more case studies prior to the exam can be highly beneficial.

Theory Explanation
Most of the solutions in this book include a theory explanation. When relevant, the theory explanation may be integrated into the body of the solution with proper details and referencing. In other situations, it may be included at the end of the solution and marked as (*).

We encourage you to read through these explanations, even if you already know the answer, and to check the references and bibliography section for any relevant sources.

Book Parts

As described earlier, the book problems are arranged in eight parts as presented in the Problems & Solutions Section. The coverage for each of the eight parts is summarized in the following paragraphs:

The *Site Characterization* part covers various surface exploration techniques, detailing the use of different apparatuses, their specifications, and applications. It also outlines methods for determining essential site characteristics, such as shear strength and rock durability and many others.

The *Soil Mechanics* part covers various soil phase relationship examples, stress/strain relationships, effective and normal stress, the construction of Mohr's circle, stress distribution and propagation, as well as permeability and relevant testing methods. This part also explains in sufficient detail the triaxial test, including the procedure and various methods of conducting this test, such as drained or undrained samples, and/or consolidated or unconsolidated conditions.

The *Construction Observation* part covers a wide range of topics related to construction operations and monitoring. It includes questions and case studies on lifting and rigging, deep excavations and dewatering, mass haul diagram theory, excavation calculations, and borrow pit volume, along with the various methods used to calculate it. Additionally, it addresses the concept of deep wells and the scour protection of piers.

The *Earthquake Engineering* part covers various topics, including shear waves, shear wave velocity, shear modulus, and shear site characterization. It also addresses seismic design categories, liquefaction and its potential, retaining wall stability during seismic events, and slope stability. Additionally, it includes methods for earthquake-relevant site characterization.

The *Earth Structures* part covers a wide range of topics on ground improvements, including interesting case studies and explanations of important conceptual topics, similar to other parts of this book. In addition to grouting techniques, slope stability methods, pavement design improvements, and the application of geosynthetics in landfills, this section aims to encourage further research. By exploring these topics, you may discover or generate more interesting case studies that could be beneficial.

The *Groundwater & Seepage* part covers a variety of topics, including dewatering, seepage in levees, groundwater flow, and its impact on nearby structures. It also discusses solutions to mitigate these impacts.

The *Problematic Soil* part delves into various types of challenging soils, including, but not limited to, collapsible, expansive, sensitive, and frost-susceptible soils. It covers their identification and provides solutions for managing these soils effectively. Additionally, this section discusses some methods for site characterization in different scenarios, offering some guidance for dealing with problematic soils.

The *Walls & Foundations* part explores the comprehensive analysis and design of a variety of earth and critical structures. This includes an in-depth examination of retaining walls under different lateral pressure conditions, anchored walls, sheet piles, and cofferdams.

For foundations, both shallow and deep types are covered extensively, whether they are single or double-shaped. The section also explores piles, providing insight into their installation methods, vertical and lateral load capacities, and various testing methods to ensure structural integrity.

Additionally, this part addresses crucial topics such as stress distribution and the resulting stress shapes within these structures. To enhance understanding, it includes several important and intriguing case studies, mirroring the approach taken throughout the rest of the book.

For ease of reference, knowledge areas covered in each of these parts are presented at each part's page break. Also, for better and quick understanding for topics covered in those parts, you can refer to the Map of Problems Presented section.

Although this book is divided into eight distinct parts, it is important to recognize that these parts are not entirely separate from each other. In fact, and due to the nature of geotechnical engineering, these parts share common themes and ideas could overlap.

About the Exam
General information

The NCEES PE exam is a rigorous exam that is administered in two sessions, a morning session, and an afternoon session. The morning session is four hours long that used to focus on broader, foundational concepts in the engineering field along with a wide range of engineering problems compared to the afternoon session. This session is replaced in 2024 in the new exam specifications with more in-depth topics that are relevant to the geotechnical civil PE exam. It is also noted that the new and beyond 2024 specifications has completely omitted any irrelevant breadth topic.

The afternoon session is also four hours long and is generally more focused on specific areas of expertise.

Both the morning and the afternoon sessions with the 2024 specifications carry a similar weight when it comes to importance and depth of coverage.

Both sessions consist of multiple-choice questions, point-and-click, drag-and-drop, or fill in the blank type. Candidates are typically required to demonstrate their ability to analyze and solve complex engineering problems during those sessions.

All in all, this exam is designed to test not only one's knowledge and technical skills, but also their ability to think critically and work under pressure.

Dissecting the Exam

The exam consists of 80 questions presented in two sessions. Candidates are given a total of 480 minutes to solve those questions. This means that each question is allotted an average time of six minutes. It is very important to keep in mind that some questions during the real exam will take only one minute to solve while others could take up to ten minutes to complete. To prepare for the exam, it is crucial to practice solving questions that have longer duration and that are more difficult, which is what this book aims to provide.

I would like to provide a reflection from my own experience sitting for the NCEES PE exam: During the exam, I was thoroughly prepared to confidently answer all questions within the allotted time, and even expected to complete the exam at a faster pace. However, I regretfully failed to study two specific chapters of a certain manual that I had assumed would have a lower probability of being included in the exam. To my dismay, two difficult questions emerged from these chapters, leaving me with no alternative but to study the relevant material from the codes and manuals provided eating away from the exam time. Consequently, I spent approximately 20-30 minutes ensuring that my answers were accurate. This unexpected event significantly impacted the remainder of my allotted exam time, and although I ultimately passed, it was an avoidable situation.

This experience is one of the reasons why I strived to provide a comprehensive coverage of all possible topics and scenarios in this book, such that exam candidates do not have to go through this experience.

How to use this Book
General

The questions presented in this book are designed with a mix of varying lengths. Some may only take a minute or two to answer, while others may require up to 10 or 15 minutes. This design has been done as such intentionally, as it reflects the format of the actual PE exam with more questions that are longer and more difficult than the exam.

By practicing with questions of varying lengths, candidates will be better prepared to manage their time during the exam. They will learn to quickly identify fewer complex questions and move through them efficiently, while also having the skills to tackle the more time-consuming questions effectively.

It is important to note that practicing only short questions, or questions that are six minutes long, may not be enough to fully prepare for the exam.

Furthermore, the variety of questions' lengths presented in this book helps keep one's mind engaged and challenged. It can be easy to become bored or disengaged when faced with a series of similar questions, but by mixing up the lengths, candidates will be forced to stay focused and adapt to the different types of questions.

Therefore, it is important to practice longer and more difficult questions in addition to the shorter ones. This will help develop one's ability to think critically, analyze complex problems, and apply their knowledge to solve them. It will also help build one's endurance and focus, which is critical for success on this exam.

Moreover, it is important to note that it is okay to spend more time on difficult questions or even shorter ones during the practice sessions. This will help identify weaknesses and areas of improvement.

As a final note, it is important to view this book as a textbook or guide as it contains, not only straightforward questions and answers, but comprehensive explanations and guidance and wide coverage to the most complex exam problems with detailed elaborations of both questions and answers. It includes theory explanation and a wide range of references that you can use at your own pace.

Which References to Own
The geotechnical civil PE exam requires you to thoroughly study 15 references, those references are presented in the relevant section in this book, and it is highly recommended to obtain a copy of each and thoroughly go through all.

References and Bibliography
In this section of the book, you will notice a subsection called References. These are the references required by the NCEES and will most likely be provided during the exam. You are required to obtain all of them and familiarize yourself with them prior to the exam.

As for the bibliography, it includes a broader list of important references either used in this book or that are important to check if you want to strengthen your knowledge in any of those areas. This bibliography list is optional; however, it is highly advisable to check at least the ones referred to in the text of some of the solutions that you will encounter in this book.

Additionally, most of the references and bibliography listed in this book, but not all, are publicly available, so feel free to browse them and see if you can obtain any of them.

2024 Exam Specifications

Effective April 2024 the exam will focus more on the depth part. This guide takes this change into account and the questions were authored with the 2024 exam specifications in mind.

The 2024 exam specifications are presented in the following two pages for ease of reference. Also, parts of the specification are presented in the Problems & Solutions relevant page breaks for each part denoting which areas are covered in that part.

SN	Knowledge Area	Expected Number of Questions
1	**Site Characterization** A. Identification, validation, and interpretation of site data and proposed site development data (e.g., aerial photography, geologic and topographic maps, GIS and geotechnical data, as-built plans, planning studies and reports) B. Subsurface exploration planning C. Exploration techniques (e.g., hollow stem auger, cased boring, mud rotary, air rotary, rock coring, sonic drilling, Cone Penetrometer Test, geophysics, and test pits) D. Sampling techniques (e.g., split-barrel sampling, thin-walled tube sampling, handling and storage) E. In situ testing (e.g., standard penetration testing, cone penetration testing, pressure meter testing, pore pressure dissipation testing, dilatometer testing, dynamic cone penetration, plate load testing, field vane shear) F. Description and classification of soils (e.g., Burmeister, Unified Soil Classification System, AASHTO, USDA, and visual/manual) G. Rock classification and characterization (e.g., recovery, rock quality designation, rock mass rating systems, weathering, discontinuity) H. Groundwater exploration, sampling, and characterization	8-12
2	**Soil Mechanics, Laboratory Testing, and Analysis** A. Soil phase relationship and index property B. Chemical, electrical, and thermal properties (e.g., non-HAZMAT) C. Stress in soil mass (e.g., total, effective) D. Stress/strain, strength E. Permeability (e.g., falling head test, constant head test, grain-size correlations)	8-12
3	**Construction Observation, Monitoring, and Quality Assurance/Quality Control and Safety** A. Earthwork (e.g., excavation, subgrade preparation, lab and field compaction, borrow studies, fill placement) B. Trench and construction safety C. Geotechnical instrumentation (e.g., inclinometer, settlement plates, piezometer, vibration monitoring) D. Temporary and permanent soil erosion and scour protection measures	6-9
4	**Earthquake Engineering and Dynamic Loads** A. Seismic site characterization B. Seismic analyses and design (e.g., liquefaction, pseudo static, earthquake loads)	5-8
5	**Earth Structures, Ground Improvement, and Pavement** A. Ground improvement (e.g., grouting, soil mixing, preconsolidation/wick drains, lightweight materials, lime/cement stabilization, rigid inclusions,	9-14

	aggregate piers) B. Geosynthetic applications (e.g., separation, strength, filtration, drainage, reinforced soil slopes, internal stability of MSE) C. Slope stability evaluation and slope stabilization D. Embankments, earth dams, and levees (e.g., stress, settlement) E. Landfills and caps (e.g., interface stability, settlements, lining systems) F. Pavement and slab-on-grade design (e.g., rigid, flexible, porous, unpaved) G. Utility design and construction	
6	**Groundwater and Seepage** A. Dewatering, seepage analysis, groundwater flow, and impact on nearby structures B. Drainage design/infiltration and seepage control	4-6
7	**Problematic Soil and Rock Conditions** A. Karst, collapsible, expansive, peat, organic, and sensitive soils B. Reactive/corrosive soils (e.g., identification, protective measures) C. Frost susceptibility D. Rock slopes and rockfalls	4-6
8	**Retaining Structures (ASD or LRFD)** A. Lateral earth pressure and load distribution B. Rigid retaining wall analysis (e.g., CIP, gravity, external stability of MSE, soil nail, crib, bin) C. Cantilevered, anchored, and braced retaining wall analysis (e.g., soldier pile and lagging, sheet pile, secant pile, tangent pile, diaphragm walls, temporary support of excavation, and beams and column elements) D. Cofferdams E. Underpinning methods and effects on adjacent infrastructure F. Ground anchors, tie-backs, soil nails, and rock anchors (e.g., design and quality control)	10-15
9	**Shallow Foundations (ASD or LRFD)** A. Bearing capacity B. Settlement, including induced stress distribution	6-9
10	**Deep Foundations (ASD or LRFD)** A. Geotechnical and structural capacity and settlement of deep foundations (e.g., driven pile, drilled shaft, micropile, helical screw piles, auger cast piles, beam/column) B. Lateral capacity and deformation of deep foundations C. Installation methods D. Static and dynamic load testing E. Integrity testing methods	10-15

Map of Problems Presented

Site Characterization

Problem 1.1 *Dam Site Rock Quality*	Problem 1.2 *Classification System (1)*
Problem 1.3 *Classification System (2)*	Problem 1.4 *Soil Shear Strength*
Problem 1.5 *Borehole Log Test Interpretation*	Problem 1.6 *Rock Durability Tests*
Problem 1.7 *Rock Mass Rating for a Tunneling Operation*	Problem 1.8 *SBT Identification from CBT Measurements*
Problem 1.9 *Sampling Collapsible Soils*	Problem 1.10 *Undrained Shear Strength for Organic Soils*
Problem 1.11 *Boreholes Locations*	Problem 1.12 *SPT Site Exploration*
Problem 1.13 *Piles Lateral Design*	Problem 1.14 *Undisturbed Soil Samples*
Problem 1.15 *Geophysical Testing Techniques and Their Use*	Problem 1.16 *USDA Soil Classification*
Problem 1.17 *Soil Stabilometer Value*	Problem 1.18 *Resilient Modulus*
Problem 1.19 *Retrieved Sample Length*	Problem 1.20 *Pore Pressure Test*

Soil Mechanics

Problem 2.1 *Settlement in Clay*	Problem 2.2 *Effective Stress Over Time*
Problem 2.3 *Soil Properties (1)*	(❋) Problem 2.4 *Soil Properties (2)*
Problem 2.5 *Soil Permeability Testing*	Problem 2.6 *Soil Moisture Content*
Problem 2.7 *Atterberg Limits*	Problem 2.8 *Principal Stresses*
Problem 2.9 *Footing Contact Pressure*	Problem 2.10 *Soil Shear Test*
Problem 2.11 *Falling Head Test*	Problem 2.12 *Buried Rigid Pipe*
Problem 2.13 *Triaxial Shear Test*	Problem 2.14 *Triaxial Failure Plane*
Problem 2.15 *Unconsolidated-Undrained Triaxial Test*	Problem 2.16 *Consolidated-Drained Triaxial Test*
Problem 2.17 *Consolidated-Undrained Triaxial Test*	Problem 2.18 *Embankment Pressure*
Problem 2.19 *Permeability Coefficient*	Problem 2.20 *Soil Sulfate Content*

Construction Observation

Problem 3.1 Soil Compaction	Problem 3.2 Trench Timber Shoring
Problem 3.3 Borrow Pit Void Ratio	Problem 3.4 Borrow Pit Moisture Content
Problem 3.5 Effective Specific Gravity	Problem 3.6 Safety Incidence Rate
Problem 3.7 Excavation Construction Safety	Problem 3.8 Noise Exposure
Problem 3.9 Crane Hands Signal	Problem 3.10 Trench Support
Problem 3.11 Soil Loss Prevention	Problem 3.12 Soil Erodibility
Problem 3.13 Excavation Slope	Problem 3.14 Construction Methods
Problem 3.15 Excavation for a Deep Pumping Station	Problem 3.16 Equipment Operations Near Power Lines
Problem 3.17 Volume of Excavation	Problem 3.18 Cut Volume
Problem 3.19 Borrow Pit Volume	Problem 3.20 Measuring Groundwater
Problem 3.21 Water Measurement	Problem 3.22 Mass Haul Diagram
Problem 3.23 Scour Protection (1)	Problem 3.24 Scour Protection (2)
Problem 3.25 Ground Measurements	

Earthquake Engineering

Problem 4.1 Physical Properties Depth	Problem 4.2 Shear Wave Velocity Test
Problem 4.3 Initial Shear Modulus	Problem 4.4 Shallow Foundation
Problem 4.5 Adjusted Base Shear	Problem 4.6 Site Classification
Problem 4.7 Seismic Design Category	Problem 4.8 Liquefaction Sensitive Soil
(✻) Problem 4.9 Liquefaction Potential	(✻) Problem 4.10 Slope Stability Pseudo Static Seismic Coefficient
Problem 4.11 Average Shear Wave Velocity	Problem 4.12 Shear Wave Velocity
(✻) Problem 4.13 Retaining Wall	Problem 4.14 Seismic Earth Pressure
Problem 4.15 Resultant of the Vertical Pseudo Static Forces	

Earth Structures

Problem 5.1 Subbase Stabilization	Problem 5.2 Cement Stabilization
(♣) Problem 5.3 Slope Stability/ Slope Safety Factor (1)	Problem 5.4 Consolidation Settlement
(♣) Problem 5.5 Base Layer Thickness	Problem 5.6 Distresses in Flexible Pavements
Problem 5.7 Consolidation Settlement	Problem 5.8 Rapid Drawdown
Problem 5.9 Slope Stability/ Slope Safety Factor (2)	Problem 5.10 Slope Stability/ Slope Safety Factor (3)
Problem 5.11 Slope Improvement	Problem 5.12 MSE Internal Stability
Problem 5.13 Highway Improvement	Problem 5.14 Landfill Design
Problem 5.15 Consolidation Wicks	Problem 5.16 Lightweight Subgrade
Problem 5.17 Vibrocompaction	Problem 5.18 Supported Embankments
Problem 5.19 Bitumen Stabilization	Problem 5.20 Edgedrain Construction

Groundwater & Seepage

Problem 6.1 Dewatering Systems	Problem 6.2 Weep Holes Flow
Problem 6.3 Dam Seepage (1)	Problem 6.4 Dam Seepage (2)
Problem 6.5 Dam Uplifting Force	Problem 6.6 Drainage Design
Problem 6.7 Heaving Safety Factor	Problem 6.8 Cofferdam Operation
Problem 6.9 Conventional Well System	Problem 6.10 Unconfined Aquifer

Problematic Soil

Problem 7.1 Collapsible Soil	Problem 7.2 Identifying Problematic Soils
Problem 7.3 Soil Sensitivity	Problem 7.4 Identifying Organic Soils
Problem 7.5 Identifying Dispersive Soils	Problem 7.6 Soil Liquefaction
Problem 7.7 Corrosive Soils	Problem 7.8 Corrosivity Scoring
Problem 7.9 Sensitive Clay	Problem 7.10 Exploratory Drilling

Problem 7.11 Sink Holes	📄 Problem 7.12 Frost Susceptible Soils
📄 Problem 7.13 Subgrade Problem Soil	📄 Problem 7.14 Frost Susceptible Soil
📄 Problem 7.15 Rock Slope Stability	

Walls & Foundations

Problem 8.1 Retaining Wall Safety Factors	Problem 8.2 Combined footing Dimensions
Problem 8.3 Distribution of Pressure under Footing	Problem 8.4 Pressure Under Footing
Problem 8.5 Foundation Settlement	Problem 8.6 Bearing Capacity for a Square Foundation
📄 Problem 8.7 Construction Operation Over a Single Footing	(⚘) Problem 8.8 Retaining Wall Safety Factor
(⚘) Problem 8.9 Ground Anchor Capacity	Problem 8.10 Pile Depth Calculation
Problem 8.11 Pile Load Carrying Capacity	Problem 8.12 Pile Cap Design
Problem 8.13 Dynamic Pile Driving	📄 Problem 8.14 Ground Improvement
Problem 8.15 MSE External Stability	Problem 8.16 Slope Rapid Drawdown Factor of Safety

(⚘) Problem 8.17 Retaining Tension Crack	Problem 8.18 Retaining Wall Loading
Problem 8.19 Average Modulus of Elasticity	Problem 8.20 Coulomb's Theory
(⚘) Problem 8.21 Gravity Wall Stress	(⚘) Problem 8.22 Rankine Various Layers
Problem 8.23 Anchored Sheet Pile	Problem 8.24 Trench Stress Evaluation
Problem 8.25 Overburden Depth	Problem 8.26 Sheet Pile Moment Diagram
Problem 8.27 Sheet Pile Cross Section	Problem 8.28 Anchors Testing
Problem 8.29 Pile Type Selection	(⚘) Problem 8.30 H Pile Depth
Problem 8.31 Pile Capacity Change	Problem 8.32 Pile Group
Problem 8.33 Pile Installation Methods	(⚘) Problem 8.34 Pile Lateral Deflection
Problem 8.35 Pile Static Testing	

(⚘) Questions flagged like this either exceed typical exam length, or could be slightly more difficult than their counterparts, but they contain crucial concepts that are worth exploring. Those questions may also contain two concepts and are worth two questions not one. We decided to combine them in one question to deliver a certain concept, idea, or area of knowledge. We highly recommend that you practice all questions regardless of their level of difficulty or length.

📄 When you see this symbol, it indicates that the question is based on either a real or a made-up case study. Since a good portion of the real exam may consist of case study questions, we have ensured there are well over 20 case studies included here. If you need additional case studies, please refer to the bibliography section.

PROBLEMS & SOLUTIONS

I. Site Characterization

II. Soil Mechanics

III. Construction Observation

IV. Earthquake Engineering

V. Earth Structures

VI. Groundwater & Seepage

VII. Problematic Soil

VIII. Walls & Foundations

Level of Difficulty (✻)

In the following sections, few questions may exceed typical exam length or may require multiple values for an answer. You may encounter similar length or level of difficulty in your exam session but not all questions are going to be that long or difficult. However, we have elected to design this book for an ultimate experience for the PE exam and we did not want to leave anything for chance.

The new exam format with an all-in depth focus, is not going to have straight forward questions, few questions are going to be of a challenging nature, and, in some instances, you may encounter some unpredictable questions. As a result, our intention in adding some challenging questions in this book is to ensure good coverage and to prepare you thoroughly to pass the exam.

For ease of reference, difficult questions, or questions that are slightly longer than average will be flagged with the symbol (✻).

We encourage you to attempt all questions regardless of their length or level of difficulty due to the important concepts they discuss and the material they attempt to deliver.

I
SITE CHARECTERIZATION

Knowledge Areas Covered

SN	Knowledge Area
1	**Site Characterization** A. Identification, validation, and interpretation of site data and proposed site development data (e.g., aerial photography, geologic and topographic maps, GIS and geotechnical data, as-built plans, planning studies and reports) B. Subsurface exploration planning C. Exploration techniques (e.g., hollow stem auger, cased boring, mud rotary, air rotary, rock coring, sonic drilling, Cone Penetrometer Test, geophysics, and test pits) D. Sampling techniques (e.g., split-barrel sampling, thin-walled tube sampling, handling and storage) E. In situ testing (e.g., standard penetration testing, cone penetration testing, pressure meter testing, pore pressure dissipation testing, dilatometer testing, dynamic cone penetration, plate load testing, field vane shear) F. Description and classification of soils (e.g., Burmeister, Unified Soil Classification System, AASHTO, USDA, and visual/manual) G. Rock classification and characterization (e.g., recovery, rock quality designation, rock mass rating systems, weathering, discontinuity) H. Groundwater exploration, sampling, and characterization

You will notice that the information needed for the questions of this part, as well as the actual exam, can often be found in more than one reference of the NCEES list of required references. The solutions offered here may highlight this, however, you are encouraged to explore all the references and skim through their chapters to discover this information. The reason for this is that one reference might contain additional insights on a subject that another does not. Knowing where to find the necessary information during the exam and being able to quickly navigate through the various references is essential.

PART I
Site Characterization

PROBLEM 1.1 *Dam Site Rock Quality*

Drilling was carried out for a dam site investigation project for a depth of 100 ft. Total length of recovered core pieces for samples > 4 in add up to 70 ft.

The below best describes the quality of rock:

(A) Very Poor

(B) Poor

(C) Fair

(D) Good

PROBLEM 1.2 *Classification System (1)*

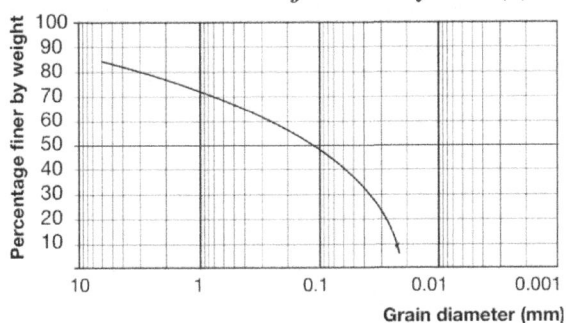

Using the Unified Soil Classification System, the above gradation sample is:

(A) GW

(B) GP

(C) SW

(D) SM

(*) Assume fines are classified as MH.

PROBLEM 1.3 *Classification System (2)*

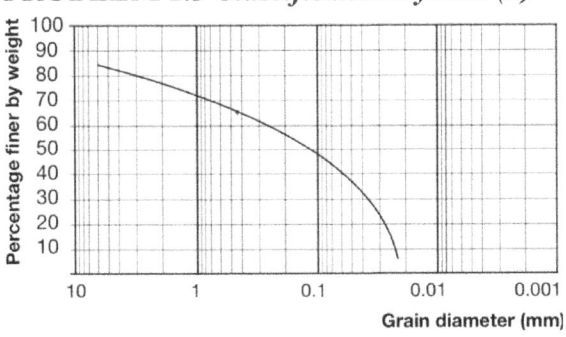

Based on the above soil sample gradation, and the following fines attributes:

- Liquid Limit (LL) = 50
- Plastic Limit (PL) = 35

Using the AASHTO classification system, the following represents the best group classification for this sample:

(A) A-7-5 (4)

(B) A-7-6 (4)

(C) A-7-4

(D) A-7-5

PROBLEM 1.4 *Soil Shear Strength*

The below is a particle distribution chart for two sands, Sand 1 (top) and Sand 2 (bottom).

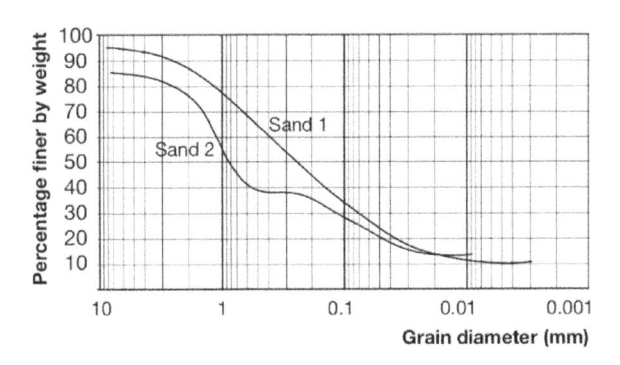

Based on this chart, the following statement is accurate during normal loading for the two sands:

(A) Sand 2 has a higher shear strength compared to Sand 1.

(B) Sand 1 has a higher shear strength compared to Sand 2.

(C) Both Sands have approximately the same shear strength.

(D) More information is required.

PROBLEM 1.5 *Borehole Log Test Interpretation*

The following page presents a borehole log for one of the project sites.

Assuming a hammer efficiency for an SPT test of $E_{eff} = 45$, the adjusted N-value for 60% efficiency incorporating an adjustment for the overburden depth for sample 3 is most nearly:

(A) 15
(B) 19
(C) 24
(D) 35

PROBLEM 1.6 *Rock Durability Tests*

The below test can be used for simple identification of degradable rocks:

(A) Jar slake test
(B) Petrographic test
(C) Rock permeability test
(D) Rock abrasion test

PROBLEM 1.7 *Rock Mass Rating for a Tunneling Operation*

A tunnel is being driven through slightly weathered sandstone with a mass of 2.5 $gram/cm^3$ at a depth of 55 $meters$ when groundwater is 10 $meters$ deep measured from the ground surface, with the layers of the sandstone at this depth formed in an unfavorable condition for tunneling.

Assume that discontinuities have slightly rough surfaces along with highly weathered walls. The strength of the intact rock is 60 MPa and the average joint frequency is 18 $joints/meter$ and the average RQD (Rock Quality Designation) has been reported as 65%.

Based on the above information, the Rock Mass Rating for this site is:

(A) 81 to 100 – very good rock
(B) 61 to 80 – Good rock
(C) 41 to 60 – Fair rock
(D) 21 to 40 – Poor rock

BOREHOLE LOG SHEET				
Depth (ft)	Groundwater Level	Graphic Log	Classification Symbol	Material Description	Samples Test Remarks (No. of SPT blows represented as x/x/x below)
0–3			CL	SANDY SILTY CLAY, medium plasticity, well graded with traces of poorly graded gravel. Density = 112 pcf	
3–7	4 ft		GW	GRAVELLY SANDY CLAY, medium plasticity, motted brown, moist to wet and fully submerged at water level at 4 ft depth. Density = 118 pcf	SPT Sample 1: 3/4/6
7–13			CL	SANDY SILTY CLAY, high plasticity, poorly graded with traces of well graded gravel. Density = 110 pcf	SPT Sample 2: 3/6/8
13–15			CL	SANDY SILTY CLAY, well graded with traces of poorly graded gravel. Density = 109 pcf	SPT Sample 3: 7/9/11

PROBLEM 1.8 *SBT Identification from CBT Measurements*

The below stress profiles have been measured using a Cone Penetration Test CPT, where the first profile to the left represents the corrected cone tip resistance (q_t) in MPa, the profile in the middle is the sleeve friction (f_s) in kPa and the last profile is the friction ratio (R_f).

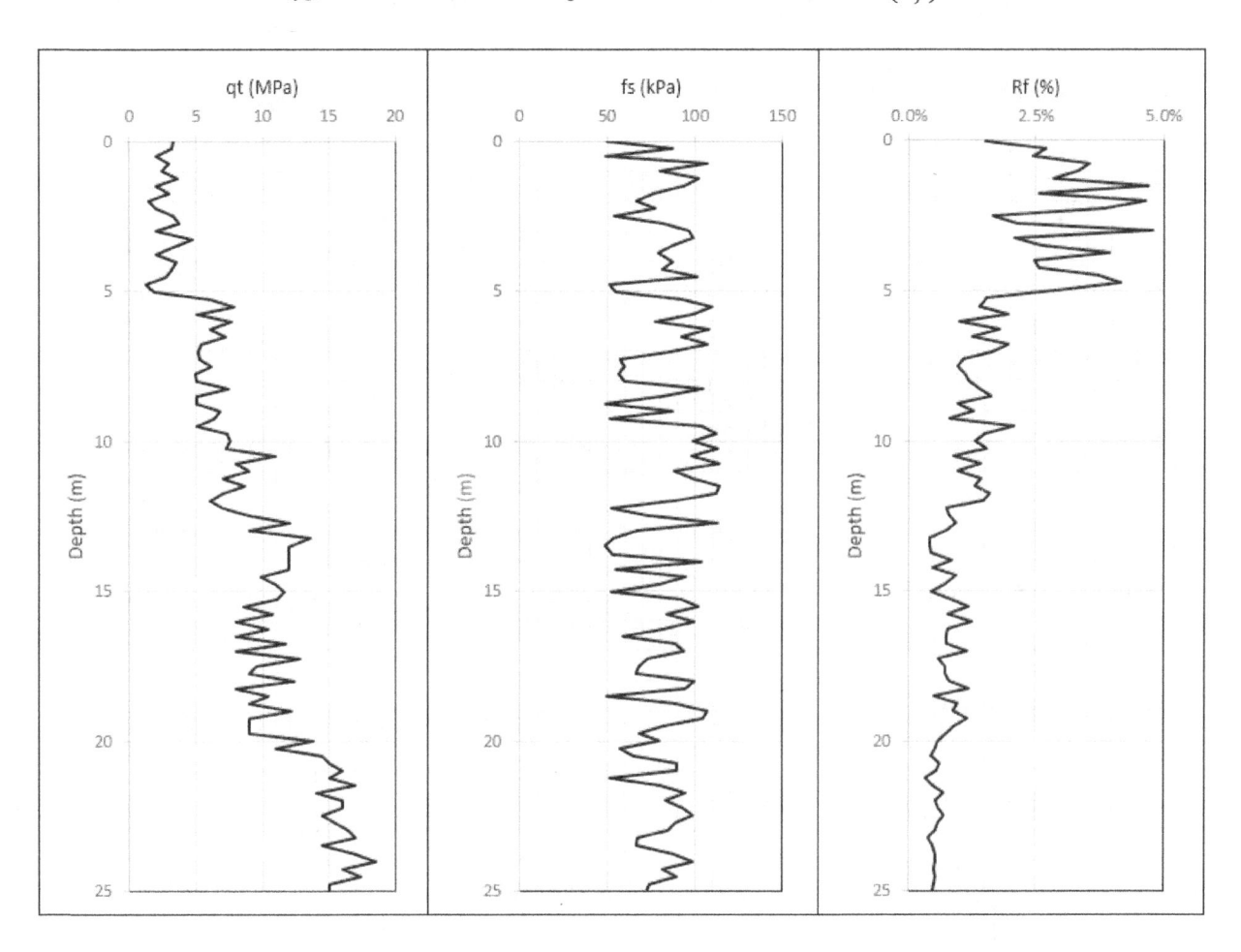

Based on the above information and in reference to the Soil Behavior Type SBT relationship with the CPT test, the type of soil found between depths 21 *to* 25 *meters* is:

(A) Sand

(B) Sensitive fine grained

(C) Clay

(D) Organic material

PROBLEM 1.9 *Sampling Collapsible Soils*
The best sampling procedure and sampler that shall be used to obtain a sample from stiff collapsible soil for direct measurements of collapse potential is:

(A) Pitcher sampler

(B) Thin-walled sampler

(C) Sonic drilling sampler

(D) Mud rotary sampler

PROBLEM 1.10 *Undrained Shear Strength for Organic Soils*
The best test that can be used to measure the undrained shear strength for non-fibrous organic soils is the following:

(A) Triaxial strength test

(B) Cone Penetration Test (CPT)

(C) Vane Shear Test (VST)

(D) Cone Penetration Test with Pore Water Pressure (CPTU)

PROBLEM 1.11 *Boreholes Locations*

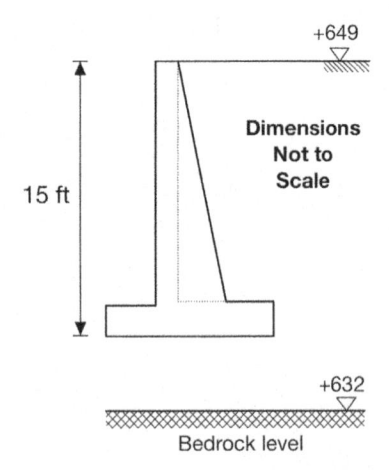

The above is a cross section of a 300 ft long, yet to be designed, retaining wall.

The minimum number, locations, and depth of the needed exploration boreholes to design this wall is/are:

(A) A borehole 6 ft deep in front of the wall below its footing level. Another one 21 ft (*) behind the wall.

(B) A borehole 16 ft (*) deep in front of the wall below its footing level. Another one 31 ft behind the wall.

(C) One borehole 21 ft (*) deep and another borehole 35 ft deep both behind the wall.

(D) One borehole 35 ft (*) deep behind the wall at mid span.

(*) Take datum level as (+649) for boreholes taken behind the wall.

PROBLEM 1.12 *SPT Site Exploration*
The best site exploration technique to be used while allowing SPT to be performed at a location with depth > 35 ft, soft clay, and with shallow groundwater, is the following:

(A) Rotary coring

(B) Hollow-stem auger boring

(C) Wash type boring

(D) Bucket auger boring

PROBLEM 1.13 *Piles Lateral Design*
The best test that can be used to obtain the design parameters required for the design of piles that are subjected to lateral loads is the following:

(A) The Dilatometer Test (DMT)

(B) The Pressuremeter Test (PMT)

(C) Vane Shear Test (VST)

(D) Cone Penetration Test with Pore Water Pressure (CPTU)

PROBLEM 1.14 *Undisturbed Soil Samples*
Select from the following the best test sampler(s) used to retrieve undisturbed soil samples:

- ☐ Split barrel
- ☐ Denison
- ☐ Large Penetration Test (LPT)
- ☐ Pitcher sampler
- ☐ Stationary Piston
- ☐ Foil Sampler
- ☐ Disc auger

PROBLEM 1.15 *Geophysical Testing Techniques and Their Use*
Match the described application for the geophysical technique to the right with the geophysical method to the left – there could be more than one application per method:

Method	Application
Seismic Refraction	Determine soil layer thickness
Spectral Analysis of Surface Waves	Determine depth to water table
DC Resistivity	Determine thickness of pavement layers
Ground Penetration Radar	Determine groundwater salinity
Electromagnetics	Determine asphalt content in asphalt concrete
Neutron Moisture Content	Determine depth to bedrock

PROBLEM 1.16 *USDA Soil Classification*
Fill out the missing information in the following table that has soil samples numbered from 1 to 3 using the below USDA Soil Texture Triangle:

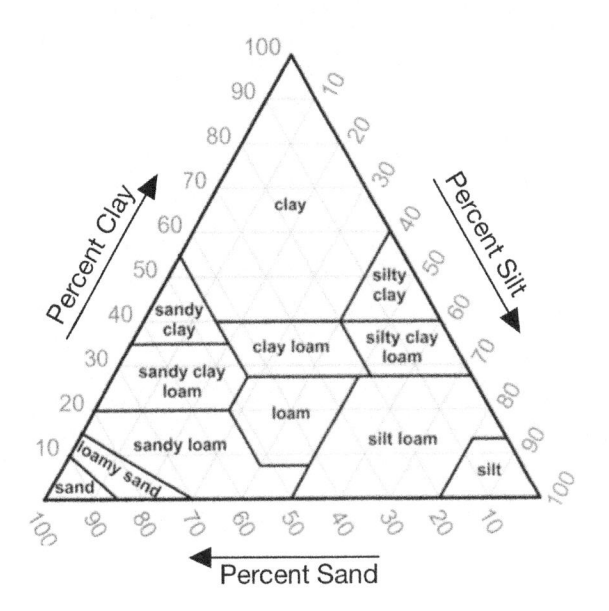

NO	% Gravel	% Sand	% Silt	% Clay	Classification
1	NA	68	22	10	??
2	20	60	18	2	??
3	NA	30	??	40	Clay Loam

ROBLEM 1.17 *Soil Stabilometer Value*

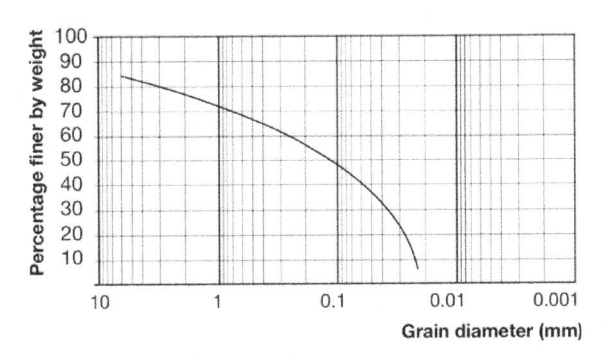

Based on the above soil sample gradation, and the following fines attributes:

- Liquid Limit (LL) = 50
- Plastic Limit (PL) = 35

The soil Stabilometer value (R-Value) is most nearly:

(A) 5

(B) 10

(C) 20

(D) 60

PROBLEM 1.18 *Resilient Modulus*
The most common method used to evaluate the resilient modulus of unbound pavement materials in the laboratory is:

(A) Conventional triaxial cell

(B) Torsional resonant column testing

(C) Simple shear test

(D) True (cubical) triaxial test

PROBLEM 1.19 *Retrieved Sample Length*
Match the described sampling method to the left with the maximum achievable retrievable soil sample length to the right:

Method	Maximum retrieved sample length
Sonic drilling	2 ft
Pitcher sampler	10 ft
Foil sampler	65 ft

PROBLEM 1.20 *Pore Pressure Test*
The soil property that is measured using the porewater dissipation test is the following:

(A) The dissipation coefficient

(B) The horizontal coefficient of consolidation

(C) The vertical coefficient of consolidation

(D) The soil rigidity index

SOLUTION 1.1

The *NCEES Handbook*, Chapter 3 Geotechnical, Section 3.7.4 Rock Classification is referred to.

Rock Quality Designation RQD is measured as follows:

$$RQD = \frac{\sum Length\ of\ Sound\ Core\ Pieces > 4\ in}{Total\ Core\ Run\ Legnth}$$

$$= \frac{70\ ft}{100\ ft}$$

$$= 70\%$$

Per the description provided in the FHA Soils and Foundation reference manual, which can be found in the *NCEES Handbook*, an *RQD* of 70% has a **Fair Quality**.

Correct Answer is (C)

SOLUTION 1.2

The following information is gathered from the gradation chart:

Nearly 45% is finer than 0.075 mm (No. 200 sieve), which means 55% is retained on this Sieve classifying the sample as either sand or gravel.

Nearly 82% is finer than 4.75 mm (No. 4 sieve), which means that 18% is retained on this sieve classifying the sample as sand.

Based on the USCS classification system found in the NCEES Handbook version 2.0 Table 3.7.2, Soils with the following:

> 50% retained on sieve No. 200

> 50% passes No. 4 sieve

Those can either be clean sands (either SW or SP), or sands with fine (either SM or SC).

Since fines are > 12% (*i.e.*, 45%), this can be classified as SM or SC.

Since the question indicated the fines are classified as MH, this sand can be classified as **SM**.

Correct Answer is (D)

SOLUTION 1.3

Using the AASHTO classification system found in the *NCEES Handbook version 2.0* Section 3.7.3, nearly 45% is finer than 0.075 mm (i.e., passing No. 200 sieve). This indicates that the sample is not granular and falls within the Silt-Clay material categories of: A-4, 5, 6 or 7.

The characteristics of fines – material finer than 0.425 mm (i.e., passing No. 40 sieve) were given as follows:

$LiquidLimit (LL) = 50$

$Plasticlimit (PL) = 35$

$Plasticity\ Index\ (PI) = 50 - 35 = 15$

Based on this, the material is classified as either A-7-5 or A-7-6.

Using the comment section of the classification table and provided that $LL - 30 = 20 > PI$, classification of the material would be that of A-7-5.

The group index GI for this category is calculated as follows:

$GI = (F - 35)\ [0.2 + 0.005\ (LL - 40)]$
$\qquad + 0.01 \times (F - 15)(PI - 10)$

$= (45 - 35)\ [0.2 + 0.005\ (50-40)]$
$\qquad + 0.01 \times (45 - 15)(15 - 10)$

$= 4$

Final classification is A-7-5 (4)

Correct Answer is (A)

SOLUTION 1.4

Generally, well graded Sands have higher friction angles compared to gap graded sands.

From the presented chart, Sand 2 seems to have gaps in its gradation around particle sizes 0.1 mm to 1.0 mm, and this will have a detrimental effect on its friction angle.

Shear strength is proportional to cohesion and to the friction angle, see below:

$$\tau = c + \sigma_n \tan\emptyset$$

- τ Shear strength
- c Total cohesion
- σ_n Normal stress
- $\emptyset$ Friction angle

It is therefore more likely that sand 1 will have a higher friction angle compared to Sand 2, which renders **Sand 1 stronger in shear during normal loading**.

Correct Answer is (B)

SOLUTION 1.5

Equations from the *NCEES Handbook version 2.0* Section 3.8.1 can be used to solve this question.

The Standard Penetration Test (SPT) is a common in-situ geotechnical test performed during boring operations. A 63.8 kg hammer is used to drive a split-spoon sampler into the soil. The sampler is hammered down by 15 cm in three attempts, each attempt measures how many blows is required to bring the earth down by 15 cm, and hence in the presented borehole log, you will find three numbers representing those three attempts.

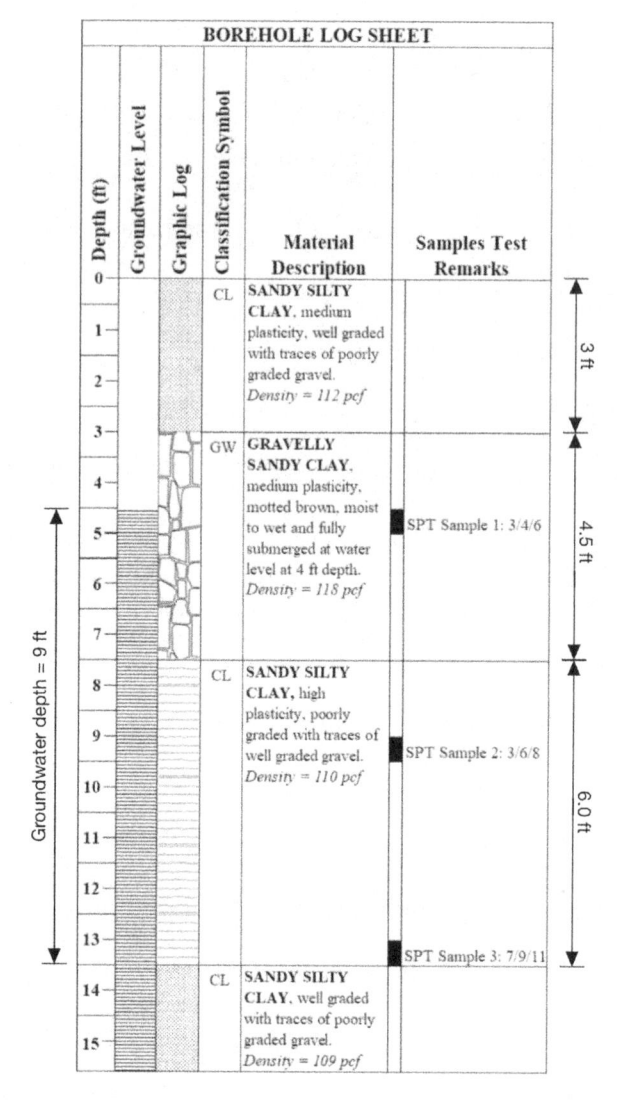

The initial set of blows that creates the first 15 cm penetration is ignored due sample disturbance from the boring operation. The second and third blows are used for analysis.

The second and third attempts are used for analysis as follows for sample 3:

$$N_{meas} = 9 + 11 = 20$$

$$N_{60} = \left(\frac{45}{60}\right) \times 20 = 15$$

The adjustment for overburden factor C_N is calculated using the effective stress P_o at the sample location (refer to Solution 2.2 for detailed steps on calculating effective stress):

$$P_o = 3 \times 112 + 4.5 \times 118 + 6.0 \times 110 - 9 \times 62.4$$
$$= 965.4 \, psf \, (0.48 \, tsf)$$

C_N can be determined by either using the graph in the same section or by solving the following equation:

$$C_N = \left[0.77 \, log_{10}\left(\frac{20}{P_o}\right)\right]$$

$$= \left[0.77 \, log_{10}\left(\frac{20}{0.48}\right)\right]$$

$$= 1.25$$

The adjusted blows for overburden depth are therefore calculated as follows:

$$(N_1)_{60} = 15 \times 1.25 \cong 19$$

Correct Answer is (B)

SOLUTION 1.6
Reference is made in this solution to the *FHWA NHI-16-072 Geotechnical Site Characterization,* Section 4.14 Rock Durability.

As mentioned toward the end of the identified section, it states that **the jar slake test is a qualitative test that is used for simple identification of degradable rocks**.

Additionally, the below is a brief description of this test along with the rest of the tests mentioned in this question:

Jar slake test: this test evaluates the stability of aggregates in water and is preferred over the traditional durability test due to its simplicity and relevance.

Petrographic test:
This test involves studying thin sections of rock samples under a microscope. It helps identify mineral composition, texture, and fabric.

Rock permeability test:
This test measures the ability of rocks to transmit fluids and it is important for understanding groundwater flow and reservoir properties.

Rock abrasion test (Los Angeles abrasion test):
This test helps determine the resistance of rock aggregates to abrasion and wear. Commonly used for assessing the durability of road aggregates.

In addition to the above, there are several other tests that are commonly used or performed on rocks, you are encouraged to check the above reference for more details on those.

Correct Answer is (A)

SOLUTION 1.7
Reference is made in this solution to the *FHWA NHI-16-072 Geotechnical Site Characterization,* Section 9.6.2 Rock Mass Rating (RMR) Classification System and Table 9-5 and 9-6 of the same section.

The RMR system uses five parameters with a sixth parameter that is used to adjust this value using the following formula:

$$RMR = \sum_{i=1}^{5} R_i$$

The parameters accompanying this equation are presented in the abovementioned tables and are described in the body of the question as follows:

Parameter 1: Strength of the intact rock:
The rating for intact rock for a $60 \, MPa$ strength is $R_1 = 7$ (Although the specific test type was not mentioned, we can infer from

the table that the uniaxial compressive strength is the parameter being referred to).

Parameter 2: Drill core quality (RQD):
For an RQD of 65%, $R_2 = 13$.

Parameter 3: Spacing of discontinuities:
With an average of 18 joints per meter, spacing is $\frac{1{,}000\ mm}{18\ joints} = 55.6$ gives an $R_3 = 5$.

Parameter 4: Conditions of discontinuities:
As described, discontinuities have slightly rough surfaces along with highly weathered walls, this generates an $R_4 = 20$.

Parameter 5: Groundwater:
There are three options under this parameter, the two that most apply are:

o Using the ratio of joint water pressure (u) to major principal stresses (σ), or,
o The general condition, knowing that the tunnel is fully submerged in water, this condition could either be wet ($R_5 = 7$) or dripping ($R_5 = 4$) or flowing ($R_5 = 0$).

Since there were enough information in this question, we can use the ratio calculation knowing that the tunnel is 55 *meters* deep, and groundwater is 10 *meters* deep – i.e., 45 *meters* submerged under water.

$$u = \gamma_{water} \times water\ depth$$
$$= 9.81\ kN/m^3 \times 45\ m$$
$$= 441.45\ kN/m^2$$

$$\sigma = \gamma_{rock} \times g \times tunnel\ depth$$
$$= 2.5\ gm/cm^3 \times 9.81\ m/s^2 \times 55\ m$$
$$= 1{,}348.9\ kN/m^2\ (*)$$

$$ratio = \frac{u}{\sigma} = \frac{441.45}{1{,}348.9} = 0.33$$

The above is equivalent to a "dripping" general condition with $R_5 = 4$.

Parameter 6: Rating:
Table 9-6 provides an $R_6 = -10$ for an unfavorable condition for tunnels:

$$RMR = R_1 + R_2 + R_3 + R_4 + R_5 + R_6$$
$$= 7 + 13 + 5 + 20 + 4 - 10$$
$$= 39\ (**)$$

This is a Poor Rock.

Correct Answer is (D)

(*) Another way of calculating principal stresses with units' conversion in mind:

$$\frac{1\ gm}{1\ cm^3} = \frac{1\ gm \times 10^{-3}\frac{kg}{gm}}{1\ cm^3 \times 10^{-6}\frac{m^3}{cm^3}} = 1{,}000\ \frac{kg}{m^3}$$

Using the NCEES handbook first chapter, stress is a unit force per area, and force is calculated as follows – use $2{,}500\ kg$ to represent the unit density per m^3 per the above:

$$F = mg$$
$$= 2{,}500\ kg \times 9.81\frac{m}{s^2}$$
$$= 24{,}525\ \frac{kg.m}{s^2}\ (i.e., 24{,}525\ N\ or\ 24.5\ kN)$$

Converting this into a force density unit

$$\gamma_{rock} = 24.5\ kN/m^3$$

The required principal stress at a depth of 45 *meters* can be calculated as follows:

$$\sigma = \frac{F}{A} = 24.53\ \frac{kN}{m^3} \times 55\ m \cong 1{,}349\ kN/m^2$$

(**) *RMR* can be used to calculate the Elastic Modulus of rock mass E_m using *FHWA NHI-06-088*, page 5-95 Eq. 5-29:

$$E_m = 145{,}000 \times \left[10^{\frac{RMR-10}{40}}\right] \cong 770{,}000\ psi$$

SOLUTION 1.8

Reference is made in this solution to the *FHWA NHI-16-072 Geotechnical Site Characterization* (*), Figure 4-14 of Section 4.17.1 SBT Identification from CPT and CPTU Measurements (original chart from Robertson, et al. 1986).

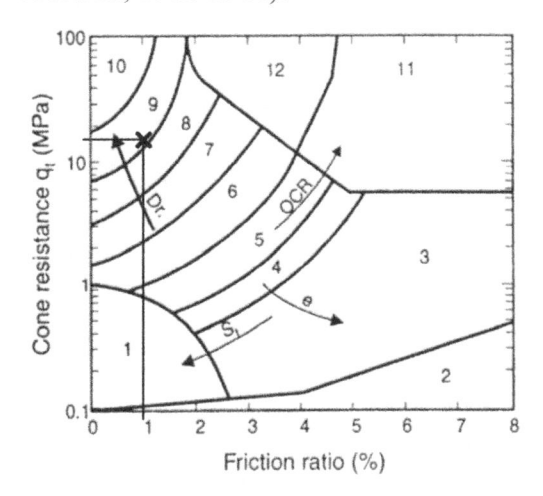

As presented in the above figure, along with the stress profiles presented in the body of the question, layers that fall between the indicated depths have a friction ratio (R_f) that falls between 0% to 1%, and a corrected cone stress (q_t) that is nearly above 14 MPa. This means, that the soil falls at SBT zone 9.

SBT Zone 9 is defined in Table 4-15 of the above reference as **Sand**.

Correct Answer is (A)

(*) This information is also available in other reference manuals, such as the *FHWA NHI-06-088 Soils and Foundations Reference Manual Volume I*, Section 3.9.7 CPT Profile Interpretation, and Figure 3-31.

SOLUTION 1.9

Reference is made in this solution to the *FHWA NHI-16-072 Geotechnical Site Characterization*, Section 5.2.4 Challenges for Subsurface Exploration in Collapsible Soils page 5-10.

As explained in the relevant section in this reference, and because the question is asking for direct measurements, the sample in this case should be undisturbed to accurately measure void ratio and interparticle bonding. A thin walled is usually used in non-stiff soil, but since the soil under question is stiff, a **Pitcher** (*) or **Denison samplers** should be used.

Correct Answer is (A)

(*) The Pitcher sampler is a subsurface sampler designed to recover undisturbed samples from formations that are too hard for thin wall like Shelby samplers or too brittle, or soft, or water sensitive for conventional core barrel type samplers. If you were not familiar with these samplers and their techniques, you are encouraged to surf the web and check how they function.

SOLUTION 1.10

Reference is made in this solution to the *FHWA NHI-16-072 Geotechnical Site Characterization*, Section 5.4.4 Shear Strength of Organic Soils and Peats.

In the relevant section of this reference, various tests are discussed for measuring undrained shear strength. Among these options, the triaxial test, cone penetration test (CPT), and cone penetration test with pore pressure measurement (CPTU) can be utilized. However, it's important to note that these methods offer less accuracy and are primarily recommended for preliminary assessments.

On the other hand, **the Vane Shear Test (VST) is effective when dealing with non-**

fibrous organic materials. Fibers in the soil can introduce fictitious tension or unrealistic shear strength, which makes the VST a preferable choice when fibers are not involved. During the VST, a "fan" equipped with vanes is inserted into the soil sample – see simple sketch below. By applying torque and spinning the fan, the test determines the undrained shear strength of the organic, non-fibrous soil.

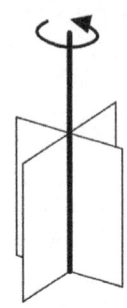

Correct Answer is (C)

SOLUTION 1.11

NCEES Handbook version 2.0 Section 3.7 provides a guideline for the minimum number of exploration points and their depth, as published by the *Federal Highway Administration* FHWA 2002 (*).

The guideline stipulates that the minimum number of boreholes required for retaining walls $> 100\,ft$ in length should be spaced $100\,ft$ to $200\,ft$ with locations alternating from the front of the wall to behind it. The depth of these exploration points should be 1 to 2 times the wall height or a minimum of $10\,ft$ below the bedrock.

Correct Answer is (B)

(*) This information is also available in *FHWA NHI-06-088 Soils and Foundations Reference Manual Volume I,* Section 3.14 Guidelines for Minimum Subsurface Exploration.

SOLUTION 1.12

Reference is made to the *FHWA NHI-06-088 Soils and Foundations Reference Manual Volume I,* Section 3.5 Boring Methods (*).

The **hollow-stem auger** has a hollow core that is initially plugged with a retractable center bit. During drilling, the center bit (the plug) is inserted into the hollow stem. However, after auguring, and when the required depth is reached, the center bit is removed to allow for the Standard Penetration Test (SPT) to be performed. The question mentioned that the soil is soft clay, which may cave in if normal auger process was performed. The hollow auger stays in place and holds the excavation augured in place until the SPT is performed.

Rotary coring, on the other hand, is primarily used for drilling through rock formations. In contrast, wash boring involves removing material using water jets, which can alter the soil properties beneath the surface, impacting subsequent SPT tests. As for the bucket auger, its maximum achievable depth is around $35\,ft$ when groundwater is present. This depth limitation makes it unsuitable for the provided scenario.

Correct Answer is (B)

(*) Ensure that you familiarize yourself with Chapter 3 of the referenced manual, which includes comparison tables for various site explorations, testing, and sampling. These tables cover topics such as:

Table 3.4: Exploratory Techniques
Table 3.5: Disturbed Soil Samplers
Table 3.6: Undisturbed Soil Samplers
Table 3.10: Factors Affecting SPT
Table 3.14: Geophysical Testing

Expect several exam questions related to these tables.

SOLUTION 1.13

Reference is made to the *FHWA NHI-06-088 Soils and Foundations Reference Manual Volume I,* Sections 3.10 and 3.11 (*).

The above reference mentions that DMT test could potentially be useful for this purpose, but this is still in the development stage. **The PMT test on the other hand is a well-developed method and can be used to determine the p-y curves needed for the design of piles subjected to lateral loads.**

Correct Answer is (B)

(*) Make sure you familiarize yourself with Chapter 3 of the referenced manual, specifically Sections 3.7 to 3.12, as you can expect several exam questions related to those sections and their relevant tables.

Additionally, and in reference to the comment provided on the page break of this part, ensure you review and familiarize yourself with this test, as well as any other test mentioned in this book or elsewhere, using various references from those listed in the NCEES exam requirements.

SOLUTION 1.14

Reference is made to the *FHWA NHI-06-088 Soils and Foundations Reference Manual Volume I,* Table 3.5 and 3.6.

- ☐ Split barrel
- ☑ Denison
- ☐ Large Penetration Test (LPT)
- ☑ Pitcher sampler
- ☑ Stationary Piston
- ☑ Foil Sampler
- ☐ Disc auger

SOLUTION 1.15

Reference is made to the *FHWA NHI-06-088 Soils and Foundations Reference Manual Volume I,* Table 3.14. You need to familiarize yourself with this table and the description it provides.

Method	Application
Seismic Refraction	1. Depth to bedrock 2. Depth to water table 3. Soil layer thickness
Spectral Analysis of Surface Waves	1. Depth to bedrock 2. Thickness of pavement layers
DC Resistivity	1. Depth to water table 2. Groundwater salinity 3. Soil layer thickness
Ground Penetration Radar	1. Depth to water table 2. Thickness of pavement layers
Electromagnetics	1. Groundwater salinity
Neutron Moisture Content	1. Estimate asphalt content in asphalt concrete

SOLUTION 1.16

The USDA Soil Texture Triangle illustrates the relative proportions of sand, silt, and clay in a soil sample. Each edge of the triangle represents a spectrum of particle sizes. Sand, with its larger particles, occupies one corner, while silt and clay occupy the other corners. The triangle's shape emerges because these three components create a spectrum between three extremes.

Along each side of the triangle, percentages ranging from 0 to 100 are marked, indicating the proportion of each component. The direction of the relevant text informs the direction of the imaginary line that shall be drawn from each edge.

Based on the above, the following procedure is implemented:

Sample No.1:
Sand, silt, and clay should sum up to 100%. Based on this, these components are plotted on the edges of the Soil Texture Triangle. The lines are extended diagonally following the direction of font in the diagram except for clay, the line is extended horizontally. The intersection of these lines indicates the soil type, in this case **Sandy Loam.**

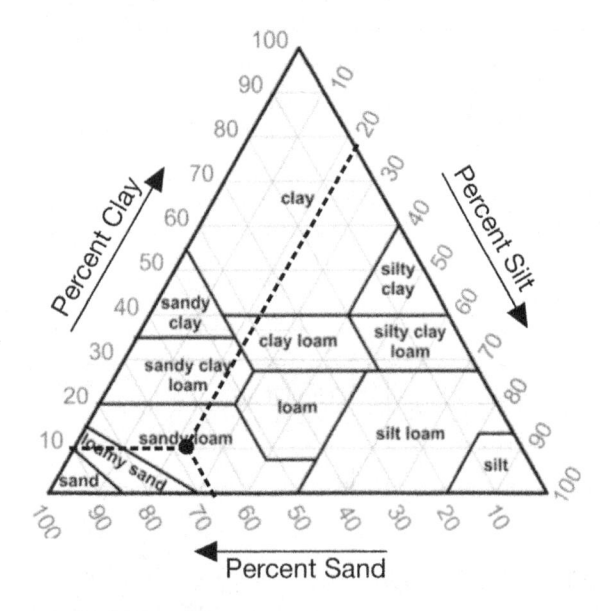

Sample No.2:
To adjust the soil gradation, first gravel is removed, then percentages are recalculated as follows: sand = 75%, silt = 22.5%, and clay = 2.5%.

Following the procedure presented in sample No. 1 the soil type in this case is **Loamy Sand** as shown below:

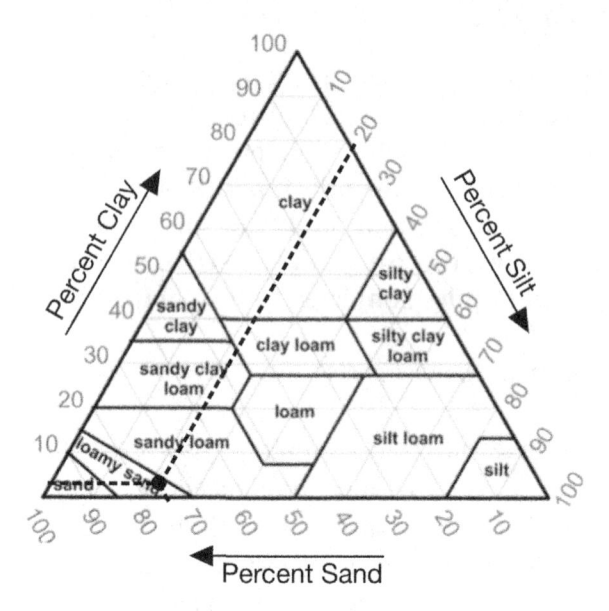

Sample No.3:
This sample follows the same logic from prior examples with silt percentage calculated as $100\% - 40\% - 30\% = 30\%$.

Based on the above, the final table looks like this:

NO	Gravel	Sand	Silt	Clay	Classification
1	NA	68	22	10	**Sandy Loam**
2	20	60	18	2	**Loamy Sand**
3	NA	30	**30**	40	Clay Loam

SOLUTION 1.17
Reference is made in this question to *FHWA NHI-05-037 Geotechnical Aspects of Pavements,* Section 5.4.3 Elastic (Resilient) Modulus, Table 5-34 (correlations).

Start with determining P_{200} which is the percent passing No. 200 Sieve (0.075 mm). Taken from the gradation graph as 45%.

From the characteristics of fines – material finer than 0.425 mm (i.e., passing No. 40 sieve) were given as follows:

$LiquidLimit(LL) = 50$

$Plasticlimit(PL) = 35$

$Plasticity\ Index\ (PI) = 50 - 35$
$= 15$

Based on this, and in reference to the abovementioned table:

$$CBR = \frac{75}{1 + 0.728\ (w\ PI)} \quad \ldots\ldots\ldots\ldots(*)$$

$$= \frac{75}{1 + 0.728\ (0.45 \times 15)}$$

$$= 12.68$$

$M_r = 2555(CBR)^{0.64}$
$= 2555(12.68)^{0.64}$
$= 12,983.4\ psi$

$R = \frac{M_r - 1,155}{555}$
$= \frac{12,983.4 - 1,155}{555}$
$= 21.3$

Correct Answer is (C)

(*) It is important to know that the originator of this equation (AASTHO) updated it in its pavement design manual for the year 2020 when FHWA (the required reference in the NCEES) for the year 2006 presents the older equation. Below is the updated equation from AASHTO:

$$CBR = \frac{75}{1 + 0.278\ (P_{200}\ PI)}$$

$$= \frac{75}{1 + 0.278\ (0.45 \times 15)}$$

$$= 26.1$$

$M_r = 2555(CBR)^{0.64}$
$= 2555(26.1)^{0.64}$
$= 20,608.2\ psi$

$R = \frac{M_r - 1,155}{555}$
$= \frac{20,608.2 - 1,155}{555}$
$= 35.1$

SOLUTION 1.18

Reference is made in this question to *FHWA NHI-05-037 Geotechnical Aspects of Pavements,* Section 5.4.3 Elastic (Resilient) Modulus, page 5-42 Unbound Material.

The **conventional triaxial cell** is the most widely used method for measuring unbound resilient modulus due to its widespread availability in laboratories. This equipment allows stress measurements similar to isolated wheel loadings. Additionally, during testing, it enables measurement of pore water pressure and strain.

Although other methods can serve the same purpose, they are less common compared to the conventional triaxial cell.

Correct Answer is (A)

SOLUTION 1.19

The below information is obtained from:

o FHWA NHI-06-088 page 3-32, Table 3-6
o UFC-3-220-10 page 62, Table 2-6

Method	Maximum retrieved sample length
Pitcher sampler	2 ft
Sonic drilling	10 ft
Foil sampler	65 ft

- UFC-3-220-10: Table 3-8 Laboratory Rock Strength Tests with ASTM Standards.
- FHWA NHI-05-037: Table 4-7 In-situ Tests for Subsurface Exploration in Pavement Design and Construction.
- FHWA NHI-05-037: Table 4-11 Subsurface Exploration-Exploratory Boring Methods.
- NCEES Handbook:
  - Page 128: Methods for Index Testing of Soils.
  - Page 130: Methods for Performance Testing of Soils.

SOLUTION 1.20

Reference is made in this solution to the *FHWA NHI-16-072 Geotechnical Site Characterization,* Section 6.13.1 CPTU Dissipation Tests.

The CPTU test is used to measure pore pressure dissipation which measures the **horizonal coefficient of consolidation (*)**.

Correct Answer is (B)

(*) You may find that the information for various tests and their properties is not centralized in a single location. This question, or variations of it, frequently appears in this examination.

Therefore, beyond reviewing all the provided references, it would be beneficial to acquaint yourself with the tables listed below as well:

- FHWA NHI-06-088 – Table 5-20 Test Methods for Rock.
- UFC-3-220-10: Table 2-18 In-situ Testing Methods used in Soil for Strength and Deformation.
- UFC-3-220-10: Table 3-5 Dynamic Tests for Soils.

II
SOIL MECHANICS

Knowledge Areas Covered

SN	Knowledge Area
2	**Soil Mechanics, Laboratory Testing, and Analysis** A. Soil phase relationship and index property B. Chemical, electrical, and thermal properties (e.g., non-HAZMAT) C. Stress in soil mass (e.g., total, effective) D. Stress/strain, strength E. Permeability (e.g., falling head test, constant head test, grain-size correlations)

PART II
Soil Mechanics

PROBLEM 2.1 *Settlement in Clay*

Time settlement in saturated clays when loaded, due to the addition of a building for example, is attributed to the following:

(A) Expulsion of clay particles

(B) The increase in effective stress of clay

(C) The deformation of clay particles

(D) All the above

PROBLEM 2.2 *Effective Stress Over Time*

The below embankment has a uniform weight of 1 ksf and was piled linearly over a period of 6 months on top of a layer of sand and clay as shown. The groundwater level is 5 ft below the embankment. The density of sand and clay layers are both 120 pcf.

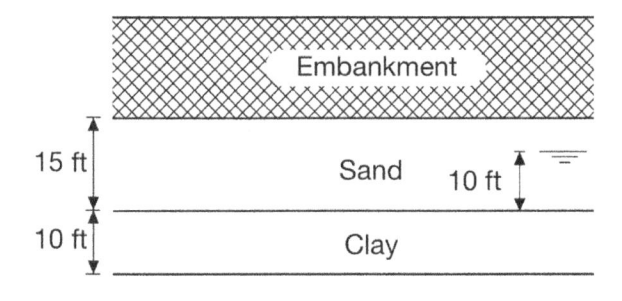

The profile that represents the change in effective stress over time at the bottom of each layer is:

(A) Profile A

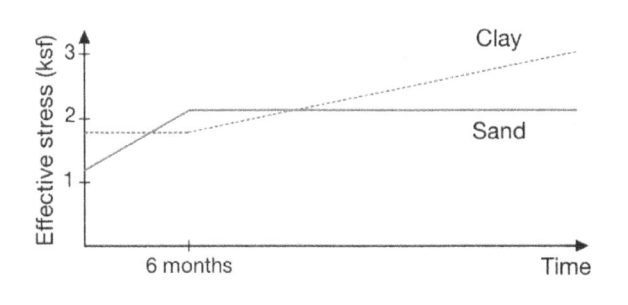

(B) Profile B

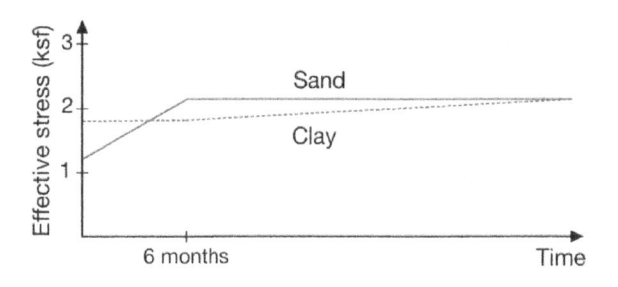

(C) Profile C

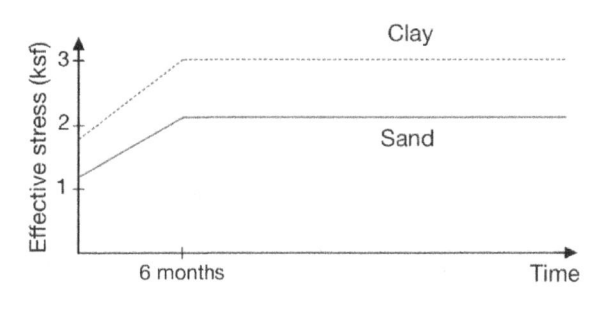

(D) Profile D

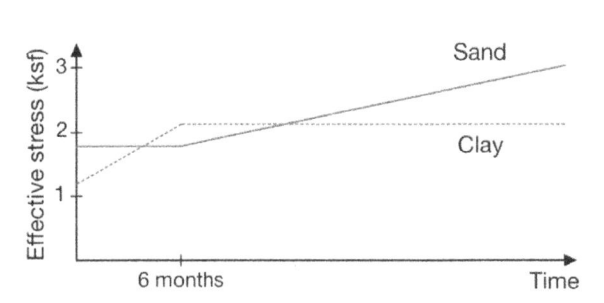

PROBLEM 2.3 *Soil Properties (1)*

The over consolidation ratio for a soil with a '0.33' normally consolidated at rest Rankine coefficient and a '0.85' over consolidated at rest coefficient is most nearly:

(A) 4.1

(B) 1.9

(C) 0.3

(D) 2.6

(✽) PROBLEM 2.4 *Soil Properties (2)*
A soil sample that has a total volume of $1\ ft^3$ and a total mass of $100\ lb$ is removed from the ground. The water content of this sample is 20% and Specific Gravity '2.7'.

Based on the above information the following attributes are as follows:

The dry density of the sample is _____

The Degree of Saturation is _____

Porosity is _____

(✽) Normally you are asked to provide one value.

PROBLEM 2.5 *Soil Permeability Testing*
The following test is recommended for use to determine the coefficient of permeability for materials with lower permeability such as silts and clays:

(A) The constant head permeameter test

(B) The Piezometric test

(C) The falling head permeameter test

(D) The flexible wall permeameter test

PROBLEM 2.6 *Soil Moisture Content*
The maximum moisture content of a soil is close to its:

(A) Liquid limit

(B) Shrinkage limit

(C) Plastic limit

(D) Plasticity index

PROBLEM 2.7 *Atterberg Limits*
A soil sample has the following Atterberg limits:
$$LL = 0.52$$
$$PL = 0.29$$
$$SL = 0.19$$

The sample's volume at its liquid limit was $15\ cm^3$. Following oven drying, the sample shrank to $9.5\ cm^3$.

Based on the above information, the specific gravity of the soil solids is most nearly:

(A) 2.52

(B) 2.63

(C) 2.71

(D) 2.83

PROBLEM 2.8 *Principal Stresses*
The soil element shown in the below sketch has normal stresses acting on it with values of $\sigma_x = 70\ psi$ and $\sigma_y = 35\ psi$. Shear stress is also acting on this element which equals to $\tau_{xy} = 25\ psi$ as shown below:

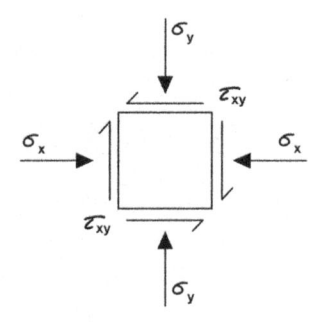

Based on this information, the magnitude of principal stresses $\sigma_{1\ (or\ max)}$ and $\sigma_{3\ (or\ min)}$ respectively are most nearly:

(A) $83\ psi, 22\ psi$

(B) $77\ psi, 28\ psi$

(C) $70\ psi, 34\ psi$

(D) $61\ psi, 18\ psi$

PROBLEM 2.9 *Footing Contact Pressure*
The below represents a rigid foundation sitting on clay with unfirm pressure on top:

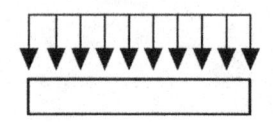

The shape of the earth contact pressure generated from the uniform pressure on top of this rigid foundation is best represented as follows:

(A) Profile A

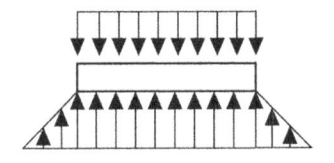

(B) Profile B

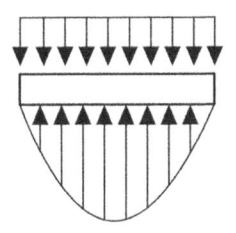

(C) Profile C

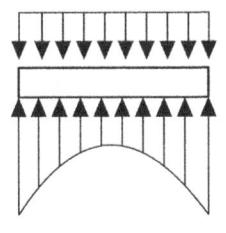

(D) Profile D

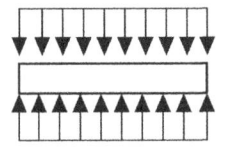

PROBLEM 2.10 *Soil Shear Test*
Three sand samples were tested using a direct shear test resulting in three distinct curves as shown below.

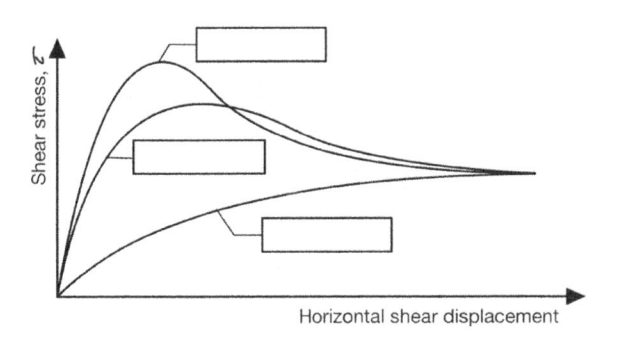

The sand samples had different densities: **dense, medium, and loose**.

Based on the shape of those curves, identify which density corresponds to which sample by filling out the boxes in the graph.

PROBLEM 2.11 *Falling Head Test*
In a falling head permeameter test, the length of the sample is 10 in and the diameter of the standpipe is 0.4 in while the diameter of the sample is 2.4 in, its hydraulic conductivity is 0.0002 ft/sec.

Based on the above information, the time it takes the hydraulic head to drop by 20 in when the hydraulic head at the beginning of the test was at 30 in is most nearly:

(A) 2 min

(B) 5 min

(C) 96 min

(D) 127 min

PROBLEM 2.12 *Buried Rigid Pipe*
The below sketch shows a 24 in dia concrete pipe buried in 6 ft deep, 4 ft wide trench backfilled with saturated sand with a total unit weight of 120 pcf. The trench is located in a highway, and it is expected that an AASHTO HS-20 truck will be used on this highway.

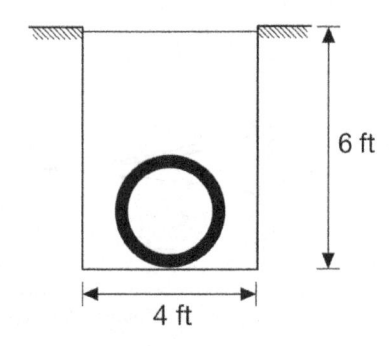

Based on the above information, the expected load per length of this pipe in this trench due to the above loading is most nearly:

(A) 2 kip/ft
(B) 5 kip/ft
(C) 7 kip/ft
(D) 11 kip/ft

PROBLEM 2.13 *Triaxial Shear Test*
Match the described triaxial shear test abbreviation to the left with the type of stress it measures to the right:

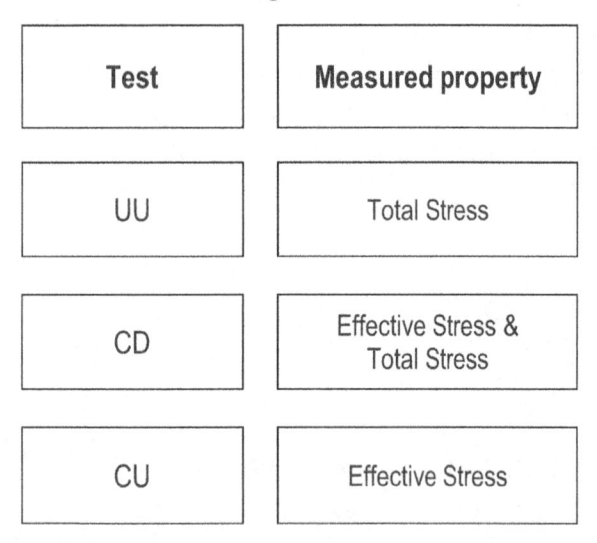

Test	Measured property
UU	Total Stress
CD	Effective Stress & Total Stress
CU	Effective Stress

PROBLEM 2.14 *Triaxial Failure Plane*
The below sketch shows a *Mohr's Circle* for the effective stress applied on a soil sample with the angle of friction showing as $35°$.

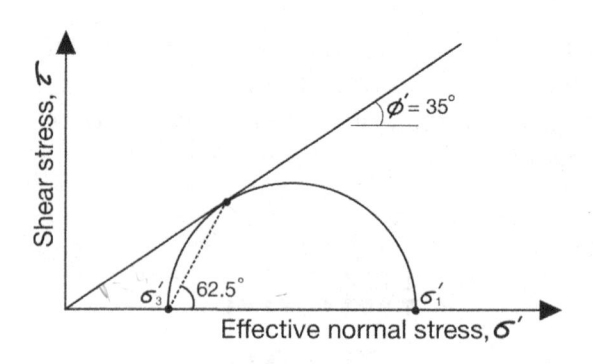

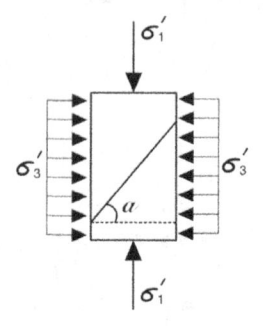

Based on the above information, the failure angle (a) for the failure plan shown on the above soil sample sketch is most nearly:

(A) $31.25°$
(B) $45°$
(C) $62.5°$
(D) $55°$

PROBLEM 2.15 *Unconsolidated-Undrained Triaxial Test*
An Unconsolidated-Undrained triaxial test was conducted on a soil sample with the chamber pressure set at $\sigma_3 = 20\ psi$.

The specimen failed at a deviator stress $\sigma_d = 55\ psi$ where the principal stress in this case $\sigma_1 = 20 + 55 = 75\ psi$.

Based on the above information, the shear stress at failure is most nearly:

(A) $22.4\ psi$
(B) $18.7\ psi$
(C) $44.8\ psi$
(D) $27.5\ psi$

PROBLEM 2.16 *Consolidated-Drained Triaxial Test*

A Consolidated-Drained triaxial test was conducted on a normally consolidated specimen with cohesion $c = 0$, and with the chamber pressure set at $\sigma_3 = 20 \, psi$.

The specimen failed at a deviator stress $\sigma_d = 55 \, psi$ where the principal stress in this case $\sigma_1 = 20 + 55 = 75 \, psi$.

Based on the above information, the effective shear stress at failure is most nearly:

(A) 22.4 psi

(B) 18.7 psi

(C) 44.8 psi

(D) 27.5 psi

PROBLEM 2.17 *Consolidated-Undrained Triaxial Test*

A Consolidated-Undrained triaxial test was conducted on a normally consolidated specimen with cohesion $c = 0$, and with the chamber pressure set at $\sigma_3 = 20 \, psi$.

The specimen failed at a deviator stress $\sigma_d = 55 \, psi$ where the principal stress in this case $\sigma_1 = 20 + 55 = 75 \, psi$. The pore pressure during the second phase of the test was $u = 10 \, psi$.

Based on the above information, the effective shear stress at failure is most nearly:

(A) 22.4 psi

(B) 18.7 psi

(C) 44.8 psi

(D) 27.5 psi

PROBLEM 2.18 *Embankment Pressure*

A load sensitive pipeline is to be placed under and to the side of the below embankment located as shown in the following sketch:

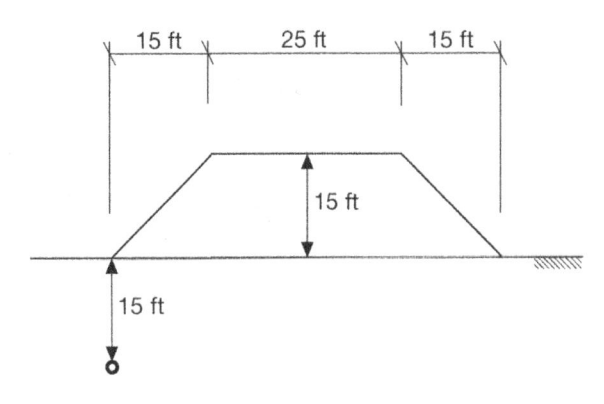

The soil density for this embankment is $120 \, pcf$.

Using the influence factor charts provided in *UFC-3-220-10 Soil Mechanics,* the expected vertical stress that will be experienced by this pipe from the embankment alone is:

(A) 430 psi

(B) 3 psi

(C) 880 psi

(D) 6 psi

PROBLEM 2.19 *Permeability Coefficient*

The following grain size distribution was collected from a gradation analysis for a certain soil sample:

$D_{10} = 0.025 \, mm$

$D_{30} = 0.037 \, mm$

$D_{60} = 0.27 \, mm$

Based on this information, the coefficient of permeability for this soil is most nearly:

(A) 25 *lugeon*

(B) 50 *lugeon*

(C) 75 *lugeon*

(D) 100 *lugeon*

PROBLEM 2.20 *Soil Sulfate Content*

The most accurate sentence from the four below options regarding soil sulfate content is the following:

(A) Generally, all sulfate rich soils can be treated with lime stabilization.

(B) Only soils with soluble sulfate concentration content > 1.0% can be treated with lime stabilization.

(C) Sulfate is normally uniformly distributed in rich sulfate soils.

(D) Mixing lime with high soil sulfate content through lime stabilization can cause soil to heave.

SOLUTION 2.1

Clay is an undrained layer, which means that when loaded, water will not drain immediately. Rather, water, due to excess pressure, will drain/get expelled slowly and over a long period of time. The slow expulsion of water from the voids between clay particles causes the layer to lose its structure which leads into its ultimate and slow settlement over time.

When clay is loaded, and due its undrained property, pore water pressure increases. Upon water expulsion, pore water pressure decreases over time resulting in a gradual long-term **increase in effective stress**.

Correct Answer is (B)

SOLUTION 2.2

The effective stress σ' is the total stress σ removed from it the pore/water pressure u. In which case, the following effective stress changes at the bottom of each layer shall occur over the indicated period of 6 months and the age of the project/embankment.

Month zero prior to placing the embankment:

$\sigma_{sand} = 120 \, pcf \times 15 \, ft$
$= 1,800 \, psf \, (1.8 \, ksf)$

$u_{sand} = 62.4 \, pcf \times 10 \, ft$
$= 624 \, psf \, (0.62 \, ksf)$

$\sigma'_{sand} = \sigma_{sand} - u_{sand}$
$= 1.8 - 0.62 \cong 1.2 \, ksf$

$\sigma_{clay} = 120 \, pcf \times (10 + 15) \, ft$
$= 3,000 \, psf \, (3.0 \, ksf)$

$u_{clay} = 62.4 \, pcf \times 20 \, ft$
$= 1,248 \, psf \, (1.25 \, ksf)$

$\sigma'_{clay} = \sigma_{clay} - u_{clay}$
$= 3.0 - 1.25 \cong 1.8 \, ksf$

Month 6 after placing the embankment:

When the embankment is placed, the sand layer will drain the excess (now pressurized) water immediately and hence no increase shall occur in the sand pore pressure. This will reflect in an increase in the effective pressure of the sand and given the loading from the embankment took place linearly over a period of 6 months, the increase in effective pressure for sand will be linear as well.

$\sigma_{sand} = 0.12 \, kcf \times 15 \, ft + 1 \, ksf$
$= 2.8 \, ksf$

$u_{sand} = 62.4 \, pcf \times 10 \, ft$
$= 624 \, psf \, (0.62 \, ksf)$

$\sigma'_{sand} = \sigma_{sand} - u_{sand}$
$= 2.8 - 0.62 = 2.2 \, ksf$

When it comes to the pore pressure of the clay layer, clay will not drain the excess (now pressurized) water right away (unlike sand). Drainage in this case shall occur over a long period of time instead. The pore pressure of the clay layer will increase due to this by the amount of the added load, and this keeps the effective stress unchanged.

$\sigma_{clay} = 0.12 \, kcf \times 25 \, ft + 1 \, ksf$
$= 4.0 \, ksf$

$u_{clay} = 0.0624 \, kcf \times 20 \, ft + 1 \, ksf$
$= 2.25 \, ksf$

$\sigma'_{clay} = \sigma_{clay} - u_{clay} = 4.0 - 2.25$
$= 1.8 \, ksf$

Over a long period of time after placing the embankment:

The sand effective stress will not change as the water has already drained from it long ago.

$$\sigma'_{sand} = \sigma_{sand} - u_{sand}$$
$$= 2.8 - 0.62$$
$$= 2.2 \, ksf$$

As for the clay layer, and over a long period of time, the excess (pressurized) water would have been drained then and this should bring the pore pressure down to:

$$u_{clay} = 0.0624 \, kcf \times 20 \, ft$$
$$= 1.25 \, ksf$$

$$\sigma'_{clay} = \sigma_{clay} - u_{clay}$$
$$= 4.0 - 1.25$$
$$= 2.8 \, ksf$$

This makes Profile A the most representative profile.

Correct Answer is (A)

SOLUTION 2.3

Reference is made to the *NCEES Handbook version 2.0*, Section 3.1.2.

At rest Rankine Coefficient for normally consolidated soils:

$$K_{o,NC} = 1 - sin\emptyset'$$
$$0.33 = 1 - sin\emptyset' \quad \rightarrow sin\emptyset' = 0.67$$

For over consolidated soils:

$$K_{o,OC} = (1 - sin\emptyset') \times OCR^{sin\emptyset'}$$
$$= K_{o,NC} \times OCR^{sin\emptyset'}$$

$$OCR = \left(\frac{K_{o,OC}}{K_{o,NC}}\right)^{\frac{1}{sin\emptyset'}}$$

$$= \left(\frac{0.85}{0.33}\right)^{\frac{1}{0.67}}$$

$$= 4.1$$

Correct Answer is (A)

(✱) SOLUTION 2.4

Referring to the *NCEES Handbook version 2.0*, Section 3.8.3:

Dry Density:

$$\gamma_d = \frac{\frac{W_t}{1+w}}{V}$$

$$= \frac{100 lb/(1+0.2)}{1 \, ft^3}$$

$$= 83.33 \, lb/ft^3$$

Degree of saturation:

$$S = \frac{w}{\left(\frac{\gamma_w}{\gamma_d} - \frac{1}{G}\right)}$$

$$= \frac{0.2}{\left(\frac{62.4}{83.33} - \frac{1}{2.7}\right)}$$

$$= 0.53$$

Porosity:

$$n = 1 - \frac{W_s}{GV\gamma_w} \left(= 1 - \frac{\gamma_d}{G\gamma_w}\right)$$

$$= 1 - \frac{83.33}{2.7 \times 62.4} = 0.51$$

SOLUTION 2.5

The **flexible wall permeameter** test is used when the tested materials' permeability is lower than $1 \times 10^{-3} cm/sec$. The specimen in this case is encased in a membrane, and with the proper amount of pressure, flow through the specimen is recorded with time.

Correct Answer is (D)

SOLUTION 2.6

The *shrinkage limit* represents the water content that corresponds to transitioning between a brittle state and a semi-solid state.

The *plastic limit* represents the water content that corresponds to transitioning between a semi-solid state and a plastic state.

The *liquid limit* represents the water content that corresponds to transitioning from a plastic state to a liquid state.

The *plasticity index* represents the range within which soil remains in its plastic state bounded with its *plasticity limit* as the lower limit, and the *liquid limit* as the upper limit to this range.

This renders the **liquid limit** the maximum water content any soil can get to.

Correct Answer is (A)

SOLUTION 2.7
Referring to the *NCEES Handbook version 2.0*, Section 3.8.3.

From the equations provided in the above section, we can use the equation relevant to the volume of solids (V_S) versus total volume (V):

$$V = V_S(1 + e)$$

Also, with the use of the equation that combines specific gravity (G), void ratio (e), water content (w), along with saturation (S), knowing that $S = 1$ for all liquid phases.

$$S \times e = w \times G$$

Based on the above, the following equations can be established:

$$e_{LL} = 0.52 \times G \quad \text{..................Equation 1}$$

$$e_{SL} = 0.19 \times G \quad \text{..................Equation 2}$$

$$V_{LL} = 15 = (1 + e_{LL})V_S \quad \text{.........Equation 3}$$

$$V_{SL} = 9.5 = (1 + e_{SL})V_S \quad \text{........Equation 4}$$

Substitute Eq.1 and Eq.2 into Eq.3 and Eq.4 respectively and then divide Eq.3 by Eq.4 as follows:

$$\frac{15}{9.5} = \frac{1 + 0.52G}{1 + 0.19G}$$

$$1.579 + 0.30G = 1 + 0.52G$$

$$0.22G = 0.579$$

$$G = 2.63$$

Correct Answer is (B)

SOLUTION 2.8
The *NCEES Handbook version 2.0*, Section 1.6.4.1 Principal Stress is referred to in this solution.

The solution requires the basic application of the following equation – see extras at the end of this solution (*):

$$\sigma = \frac{\sigma_x + \sigma_y}{2} \mp \sqrt{\left(\frac{\sigma_x - \sigma_y}{2}\right)^2 + \tau_{xy}^2}$$

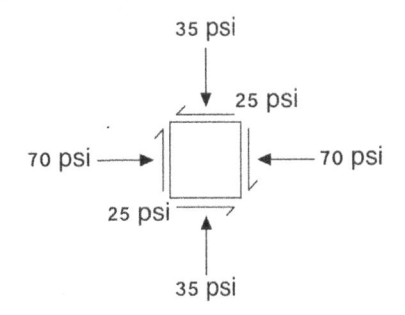

$$\sigma_1 = \frac{70 + 35}{2} + \sqrt{\left(\frac{70 - 35}{2}\right)^2 + 25^2} = 83 \; psi$$

$$\sigma_3 = \frac{70 + 35}{2} - \sqrt{\left(\frac{70 - 35}{2}\right)^2 + 25^2} = 22 \; psi$$

Correct Answer is (A)

(*) Construction of *Mohr's Circle*:

The following method can be used to find the angle that generates the above calculated principal stresses.

Initially begin with determining the location of the center of the circle 'C' and its radius 'R' as follows:

$$C = \frac{\sigma_x + \sigma_y}{2}$$
$$= \frac{70 + 35}{2}$$
$$= 52.5 \; psi$$

$$R = \sqrt{\left(\frac{\sigma_x - \sigma_y}{2}\right)^2 + \tau_{xy}^2}$$
$$= \sqrt{\left(\frac{70 - 35}{2}\right)^2 + 25^2}$$
$$= 30.5 \; psi$$

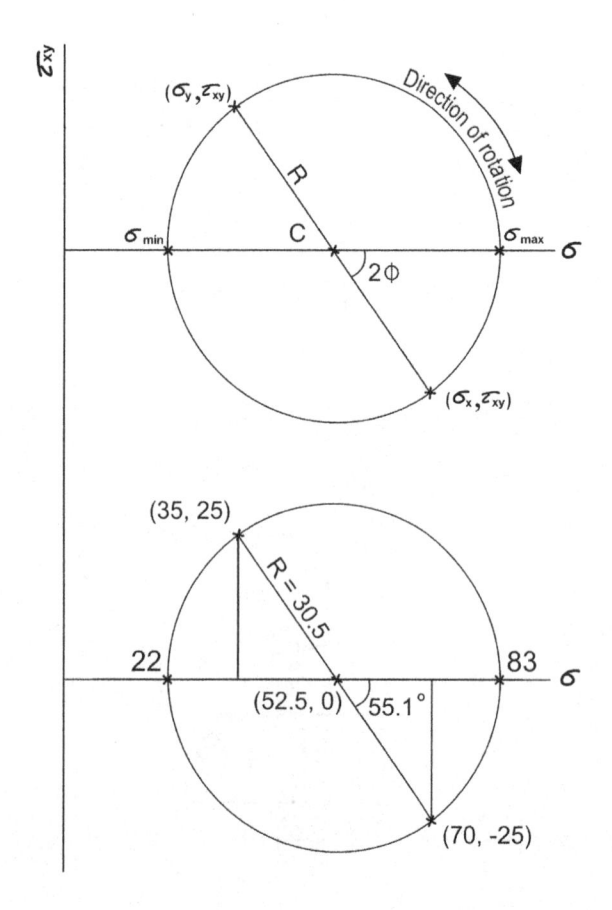

Mohr's Circle can be constructed in reference to the *NCEES Handbook version 2.0* Section 1.6.4.2.

The maximum and minimum ordinates can be determined as follows:

$$\sigma_{max} = C + R$$
$$= 52.5 + 30.5$$
$$= 83 \; psi$$

$$\sigma_{min} = C - R$$
$$= 52.5 - 30.5$$
$$= 22 \; psi$$

The angle which the maximum stress occurs at is (ϕ) and can be determined using basic trigonometry as follows:

$$2\phi = \sin^{-1}\left(\frac{25}{30.5}\right)$$
$$= 55.1°$$

$$\phi = 27.5°$$

The direction of rotation showing in the *Mohr's Circle* sketch indicates the direction that takes the soil element from the principal stresses to the stresses given in the question. In order to evaluate the direction to rotate this element to achieve the required principal stresses, rotation is opposite to that – it also depends on the negative sign you decide to assign to the shear stresses. See below:

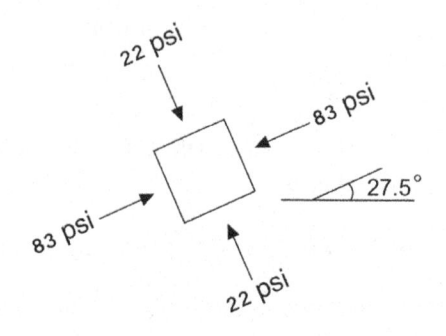

SOLUTION 2.9

Reference is made to the *UFC-3-220-10 Soil Mechanics,* Section 4-2.1.5 Applied Loads.

When a rigid foundation is subjected to a uniform load, it experiences a uniform settlement. As this foundation rests on clay, the pore pressure within the clay counteracts the load. This resistance causes the pressure to disperse laterally towards the periphery, resulting in a contact pressure that resembles profile C.

Correct Answer is (C)

SOLUTION 2.10

The below graph has been filled out with the correct answers – explanation follows:

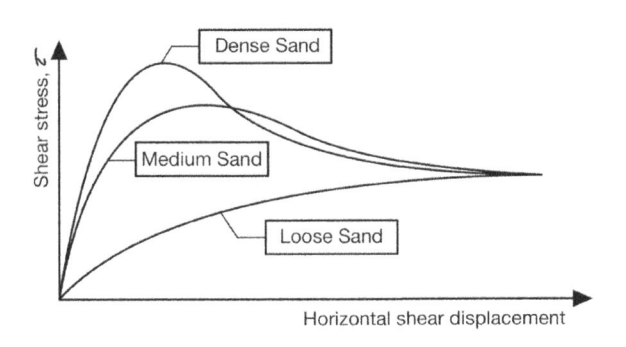

In dense sand, shear stress behavior during loading is influenced by the sand's structure and response to stress. Initially, dense sand, with its closely packed particles and limited void spaces, exhibits high interparticle friction, resulting in higher resistance. Upon reaching peak stress, the particles begin to rearrange, reducing effective contact and causing shear stress to decrease as deformation continues. Dense sand may initially reduce in volume when loading starts, then expands as loading continues due to particle rearrangement.

In loose sand, the shear stress behavior differs due to its initial structure. At the start of loading, the loose nature and higher void spaces result in low resistance. As particles rearrange, resistance increases, peaking at a lower stress level compared to dense sand. Loose sand reduces in volume as particles rearrange.

SOLUTION 2.11

Reference is made to the *FHWA NHI-16-072 Geotechnical Site Characterization,* Section 10.2.3 Fundamental Means for Measurement of Hydraulic Conductivity.

The same information is provided by the *NCEES Handbook version 2.0* page 140.

The following equation can be used to solve this question:

$$k = \frac{aL}{A\Delta t} \ln\left(\frac{h_1}{h_2}\right)$$

$$\rightarrow \Delta t = \frac{aL}{Ak} \ln\left(\frac{h_1}{h_2}\right)$$

Where:

a is the standpipe cross-sectional area $= 0.126\ in^2$

A is the sample cross-sectional area $= 4.52\ in^2$

$h_1 = 30\ in$

$h_2 = 10\ in$

$$\Delta t = \frac{0.126\ in^2 \times 10\ in}{4.52\ in^2 \times 0.0002\ ft/sec \left(\times \frac{12\ in}{1\ ft}\right)} \times \ln\left(\frac{30}{10}\right)$$

$$= 127.6\ sec\ (2.1\ min)$$

Correct Answer is (A)

SOLUTION 2.12

Reference is made to the *UFC-3-220-10 Soil Mechanics,* Section 4-4.2 Vertical Loads on Rigid Pipe.

Use Equations 4-6 and 4-7 along with Tables 4-3, 4-4 and 4-5 of the abovementioned reference as follows:

Calculate Dead Load:

$$W_d = C_d \gamma_t B_d^2$$

Where:

$$C_d = \frac{1 - e^{-Ku'(H/B_d)}}{2Ku'}$$

H is depth of trench above the pipe = $4\ ft$

$K = 0.33$ per Table 4-3

$u' = 0.5$ per Table 4-3

$Ku' = 0.17$

$\gamma_t = 120\ pcf$ per Table 4-3 (which is also given in the question)

B_d is the width of the trench = $4\ ft$

$$C_d = \frac{1 - e^{-0.17(4/4)}}{2 \times 0.17} = 0.46$$

$$W_d = 0.46 \times 120\ lb/ft^3 \times (4\ ft)^2$$
$$= 883.2\ lb/ft$$

Calculate Live Load:

Use values provided from Table 4-4 (value of '1' as substituted in the below equation) and Table 4-5 (value of 2.78 *psi* as substituted in the below equation):

$$W_l = 1 \times 2.78 \frac{lb}{in^2} \times 24\ in$$
$$= 66.72\ lb/in\ (800.6\ lb/ft)$$

Per Section 4-4.2.2, Table 4-5 has an impact factor of 1.5 included in it.

Calculate total load:

$$W_d + W_l = 1{,}683.8\ lb/ft\ (1.7\ kip/ft)$$

Correct Answer is (A)

SOLUTION 2.13

This solution references *FHWA NHI-16-072 Geotechnical Site Characterization,* Chapter 7 Measurement and Interpretation of Shear Strength Properties of Soil.

The triaxial shear test uses an undisturbed soil sample that is fully saturated before testing. The sample is placed on a pedestal within a triaxial cell, capped at the top, and encased in an impermeable latex membrane. This membrane isolates the soil from the cell-filling fluid, as shown in the below sketch:

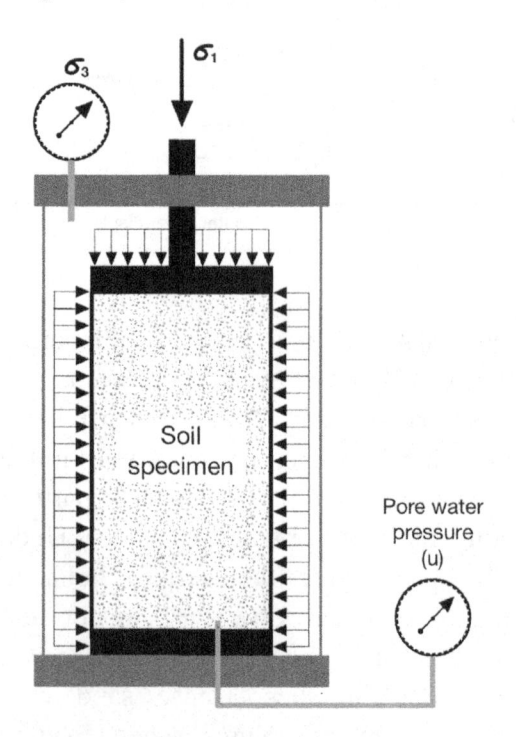

Two gauges are attached to the cell:
- The minor stress gauge (σ_3): Measures the fluid pressure inside the cell.

- The pore pressure gauge (u): Measures the pressure within the soil specimen.

Moreover, there is a valve within the cell specimen that regulates soil drainage. When open, the test is considered drained (D), resulting in zero pore pressure. When closed, test is undrained (U) resulting in excess pore water pressure.

There are two phases for this test:

1. The consolidation phase: Cell pressure is increased by the surrounding fluid by an amount (σ_3) – the minor principal stress. This increase allows the sample to consolidate if the valve is open.

2. The shear phase: A piston applies the major principal stress (σ_1) to the top of the specimen. This stress is increased gradually until the specimen fails.

This setup allows three tests to be performed (*) differentiated by two-letters that indicate consolidation/drainage conditions:

Unconsolidated-Undrained (UU):
The drainage valve remains closed throughout the test, measuring <u>total stress</u> without any pore pressure. This is a rapid test because it bypasses the need for consolidation.

Consolidated-Drained (CD):
The drainage valve stays open, allowing for consolidation with zero pore water pressure, thus measuring the <u>effective stress</u> without excess pore water pressure accumulation.

Consolidated-Undrained (CU):
The valve is open during consolidation but closed during the shear phase. This test measures both <u>total and effective stresses</u>. After drainage, pore pressure is measured and subtracted to determine the effective stress from the total stress. See below:

Test	Measured property
UU	Total Stress
CD	Effective Stress
CU	Total Stress & Effective Stress

(*) A fourth test can be performed which is the Undrained Compression (UC) test. This is a normal shear strength test with the valve open (undrained) and no minor stresses applied.

SOLUTION 2.14
This solution references *UFC-3-220-10 Soil Mechanics,* Figure 4-3 Mohr Circle Relationships. This figure is replicated below with the information provided in the question.

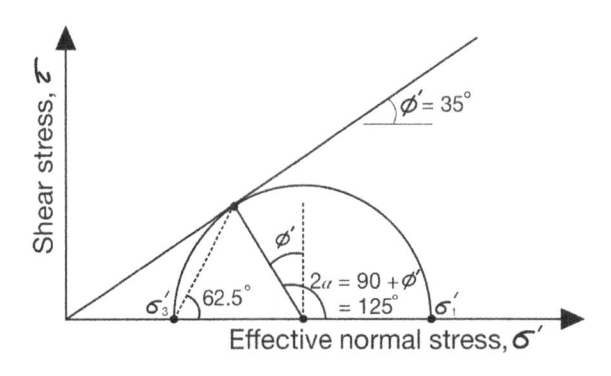

Based on the above calculation made on the sketch, the angle of failure $a = \frac{125}{2} = 62.5^o$

Correct Answer is (C)

SOLUTION 2.15

This solution references Solution 2.13 here and the explanation provided for the Unconsolidated-Undrained test.

From this explanation, it is discussed that this test measures the total stress, in which case the sample will always fail at the deviator stress $\sigma_d = 55\ psi$.

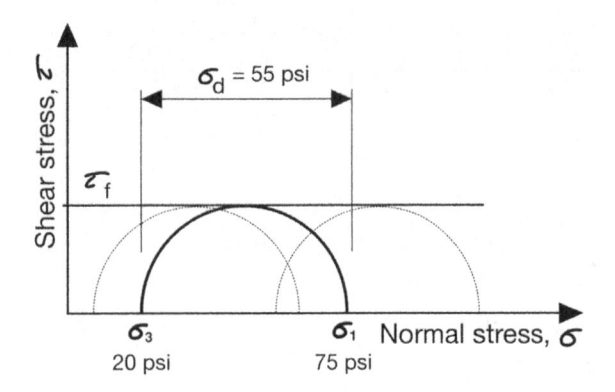

Mohr's Circle for this case is constructed as shown above. Similar hypothetical cases are added in dotted lines/circles showing the sample always fails at the same σ_d.

Based on this, shear stress at failure τ_f is calculated as follows:

$$\tau_f = \frac{55}{2} = 27.5\ psi$$

Correct Answer is (D)

SOLUTION 2.16

This solution references Solution 2.13 here and the explanation provided for the Consolidated-Drained test. This explanation clarifies that the test measures the effective stress in the soil sample, which is achieved by maintaining drainage throughout the test procedure. Additionally, the deviator stress (σ_d') is applied gradually to prevent the contribution of residual pore pressure, thus ensuring the specimen remains drained.

Contrary to the *Mohr's Circle* for the UU test, which yields a horizontal failure envelope, the failure envelope in this scenario is inclined and intersects with the origin, indicating zero cohesion as given in the question. See below:

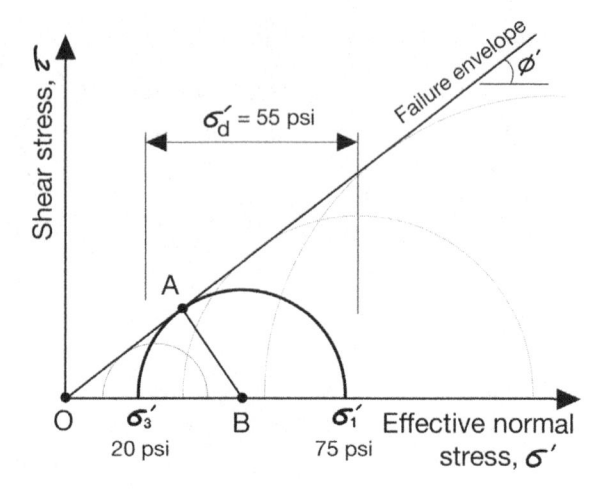

The inclination of the failure envelope is attributed to conducting multiple tests with varying confining stresses (σ_3) represented by light dotted grey circles in the above sketch. During these tests, the specimen fails at different shear stress levels, unlike the UU test. Connecting these failure points generates an inclined failure envelope, delineating an *impossible zone* above it. This zone signifies that no stress state can exist above the envelope for any given confining stress.

Trigonometry is used to compute the effective friction angle ($\emptyset'$) as follows:

$$\emptyset' = sin^-\left(\frac{line\ AB}{line\ OB}\right)$$

$$= sin^-\left(\frac{\sigma_d'/2}{\sigma_3' + \sigma_d'/2}\right)$$

$$= sin^-\left(\frac{27.5}{20+27.5}\right)$$

$$= 35.4°$$

Based on this, shear stress at failure $\tau_{f,A}$ is calculated using trigonometry in reference to the below sketch:

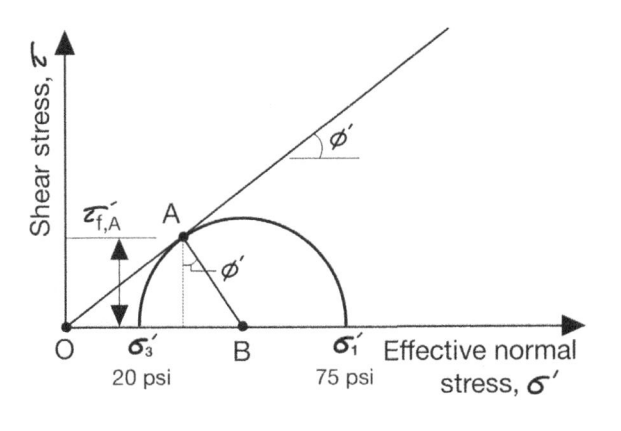

$$\tau'_{f,A} = \left(\frac{\sigma_{d'}}{2}\right) cos(\emptyset')$$

$$= 27.5 \times cos(35.4°)$$

$$= 22.4 \; psi$$

Correct Answer is (A)

SOLUTION 2.17

This solution uses the method provided in Solution 2.16 here with few adjustments. It also refers to the explanation provided for the Consolidated-Undrained test of Solution 2.13.

The CU test measures the total and the effective stress in the sample. Phase I of the test starts with the consolidation phase where the drain is open, then phase II continues with the drain shut while applying the deviator stress (σ_d') thus creating pore pressure (u). Pore pressure is then deducted (if positive) or added (if negative) from/to the minor stress (σ_3) and the principal stress (σ_1) generating the relevant effective stresses and effective friction angle ($\emptyset'$). See sketch below:

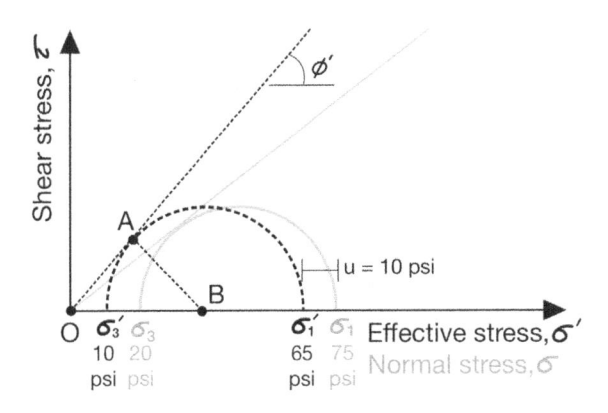

Based on this understanding, the process provided in Solution 2.16 is repeated here:

$$\emptyset' = sin^{-}\left(\frac{\sigma_d'/2}{\sigma_3'+\sigma_d'/2}\right)$$

$$= sin^{-}\left(\frac{27.5}{10+27.5}\right)$$

$$= 47.18°$$

$$\tau'_{f,A} = \left(\frac{\sigma_{d'}}{2}\right) cos(\emptyset')$$

$$= 27.5 \times cos(47.18°)$$

$$= 18.7 \; psi$$

Correct Answer is (B)

SOLUTION 2.18

This solution references *UFC-3-220-10 Soil Mechanics,* Figure 4-7 Influence Factors for Embankment Loading, also Section 4-3.2.4 Superposition, which indicates that the principle of superposition can be applied in this case.

Taking superposition into account, the below grey section marked as '1' is added to the embankment so we can use the shape provided in Figure 4-7 which will be deducted from it at the final stage.

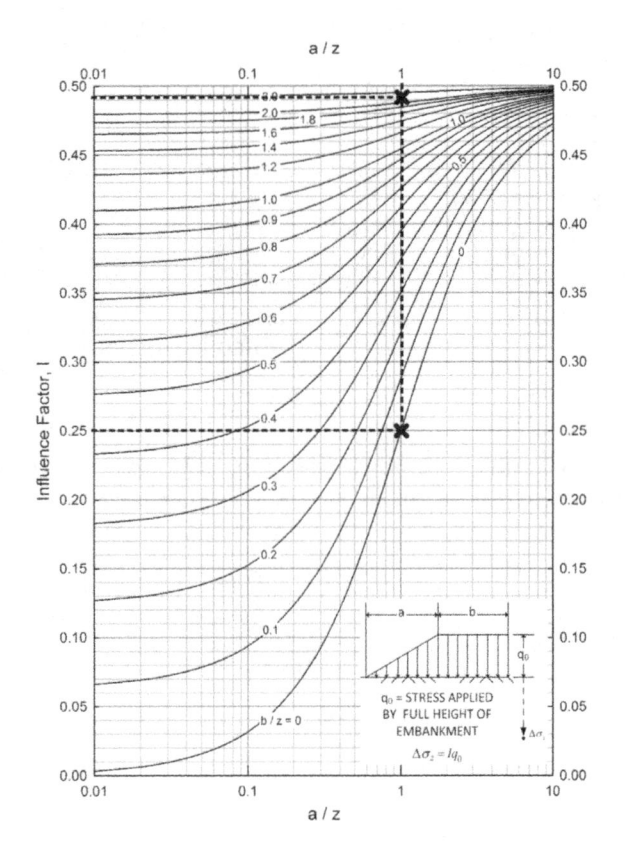

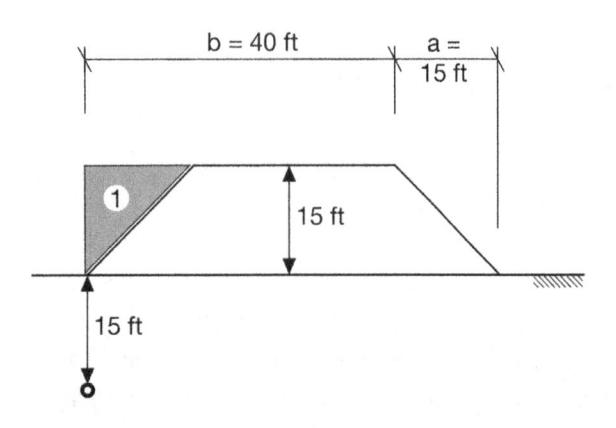

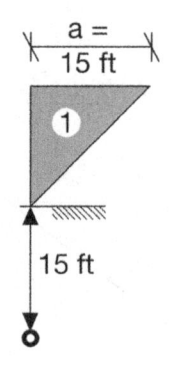

For the above section, $b/z = 0$, $a/z = 1$. From the referenced chart, copied here for ease of reference, $I_1 = 0.25$.

Based on the above, the vertical stress at point A is calculated as follows:

$$\Delta \sigma_A = I q_o$$

$$= (I - I_1) \times (h \times \gamma)$$

$$= (0.49 - 0.25) \times (15\ ft \times 120\ pcf)$$

$$= 432\ psf\ (3\ psi)$$

Correct Answer is (B)

SOLUTION 2.19

This solution references the *NCEES Handbook version 2.0,* Section 3.8.5.2 Hazen's Equation for Permeability, and *UFC-3-220-10 Soil Mechanics,* Section 8-7.2 Correlations for Coarse-Grained Soils, where coefficient C is better defined in the latter reference.

Additionally, it is recommended to review the 'Correlations' section in UFC-3-220-10 to gain a deeper understanding of the various possible correlations, which will aid in better exam preparation.

Based on the above, the hydraulic conductivity (i.e., the coefficient of permeability) is computed as follows:

For the above amended section, $b/z = 2.67$, $a/z = 1$. From the referenced chart, copied her for ease of reference, $I = 0.49$.

In a similar fashion, and as shown below, the additional section shall be deducted from the final calculation.

$$k = C(D_{10})^2$$

Where C in this equation is defined in UFC-3-220-10 Section 8-7.2 as an empirical coefficient usually taken to be $1\ cm/s/mm^2$.

$$k = 1 \times (0.025)^2$$
$$= 6.25 \times 10^{-4}\ cm/sec$$
$$= 48.1\ lugeon$$

Section 3.8.5.3 from the *NCEES handbook version 2.0* is used for the conversion to *lugeon*.

Correct Answer is (B)

SOLUTION 2.20
Reference is made to the *FHWA NHI-16-072 Geotechnical Site Characterization,* Section 5.12 High Sulfate Soils.

As mentioned in the above reference, the calcium lime used in cement stabilization reacts with natural sulfates to produce expansive minerals, which can cause heaving and pose potential hazards to transportation infrastructure. Additionally, it states that sulfates are rarely uniformly distributed throughout soil profiles.

Based on this information, the only true statement here is option (D).

Correct Answer is (D)

PART II
Soil Mechanics

III
CONSTRUCTION OBSERVATION

Knowledge Areas Covered

SN	Knowledge Area
3	**Construction Observation, Monitoring, and Quality Assurance/Quality Control and Safety** A. Earthwork (e.g., excavation, subgrade preparation, lab and field compaction, borrow studies, fill placement) B. Trench and construction safety C. Geotechnical instrumentation (e.g., inclinometer, settlement plates, piezometer, vibration monitoring) D. Temporary and permanent soil erosion and scour protection measures

PART III
Const. Observation

PROBLEM 3.1 *Soil Compaction*
A proctor test was performed on *four* soil samples from the same batch using a proctor standard mold (*).

The weight of the moist samples after applying the test blows were: 4.32 lb, 3.92 lb, 3.92 lb and 4.36 lb, and their moisture content 11.6%, 17%, 8.8% and 14.8% respectively.

The optimum moisture content for this batch is most nearly:

(A) 14.0%

(B) 11.3%

(C) 12.5%

(D) 14.8%

(*) The volume of the mold is $1/30\ ft^3$.

PROBLEM 3.2 *Trench Timber Shoring*
A $20\ ft$ deep, $15\ ft$ wide, trench is dug in cohesive soil with unconfined compressive strength of $45\ kPa$ ($\sim 0.9\ ksf$ or $\sim 0.45\ tsf$) requires the minimum following shoring system to prevent it from caving:

(A) 10×10 cross braces, $6\ ft$ horizontal spacing and $5\ ft$ vertical spacing.

(B) 8×8 cross braces, $6\ ft$ horizontal spacing and $5\ ft$ vertical spacing.

(C) 6×6 cross braces, $8\ ft$ horizontal spacing and $5\ ft$ vertical spacing.

(D) 4×4 cross braces, $8\ ft$ horizontal spacing and $5\ ft$ vertical spacing.

PROBLEM 3.3 *Borrow Pit Void Ratio*
A road contractor signed an agreement with an entity that owns a borrow pit to procure fill material at a unit rate of $2.0 per cubic yard. The borrow pit has a void ratio of 0.85 and a specific gravity of $G_S = 2.7$.

The contractor wishes to construct an embankment with $200{,}000\ ft^3$ from this borrow pit with a dry density at this embankment of $100\ lb/ft^3$. The cost of the fill material excluding transportation or waste that is incurred by the contractor in this case is most nearly:

(A) 13,453

(B) $14,815

(C) $16,314

(D) $18,519

PROBLEM 3.4 *Borrow Pit Moisture Content*
A borrow pit that is characterized with a moisture content of 8% and a total density of $110\ lb/ft^3$ provides fill material to an embankment that requires 14% moisture to achieve a dry density of $114.5\ lb/ft^3$.

The amount of water that needs to be added for each cubic yard of this fill in order to achieve the embankment's required dry density during compaction is most nearly:

(A) 10 *gallons*

(B) 20 *gallons*

(C) 30 *gallons*

(D) 40 *gallons*

PROBLEM 3.5 *Effective Specific Gravity*
For an asphalt mix, the percent by weight of stones is 95.2% and the maximum specific gravity for the paving mixture with no air voids is 2.55, the specific gravity for asphalt itself is 1.05.

Using this information, the effective specific gravity for the aggregates is most nearly:

(A) 2.55

(B) 2.35

(C) 1.02

(D) 2.75

PROBLEM 3.6 *Safety Incidence Rate*

The number of safety incidents on a 2 *year* construction project with 150 workers spending 260 *hrs* per month along with an absence rate of 15% is 250.

Half the workers moved into a new 1.5 *year* project with the same working conditions.

Assuming 25% improvement on safety, the expected number of incidents in the second project is most nearly:

(A) 117

(B) 70

(C) 150

(D) 120

PROBLEM 3.7 *Excavation Construction Safety*

A mobile crane is required to lower pipes into the trench shown below. The distance from the center of rotation of the boom to the edge of excavation is 7.5 *ft*. Also, the center of rotation is 7 *ft* above ground.

Drawing Not To Scale

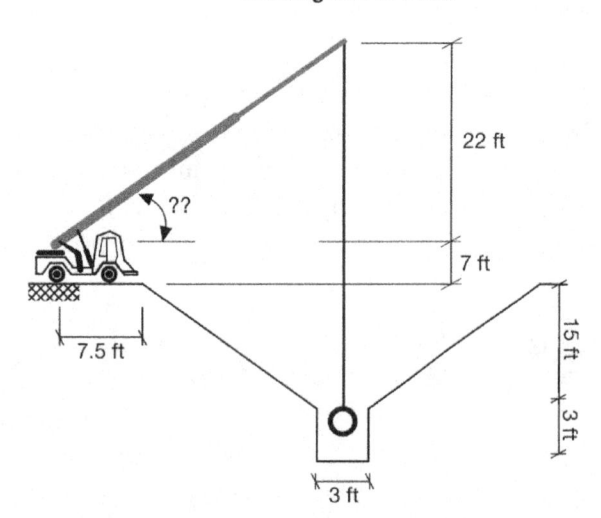

Knowing that the soil type is C per OSHA's definition of soil types, the boom angle is most nearly:

(A) 42.6°

(B) 34.9°

(C) 32.3°

(D) 44.0°

PROBLEM 3.8 *Noise Exposure*

Workers are exposed to the following noise during their *eight* hours typical working day in a construction site:

Noise Level dBA	Exposure Time hr
80	2.0
85	2.0
90	3.0
95	1.0

The amount of noise dose those workers are exposed to during their typical working day is most nearly:

(A) 350 dBA

(B) 86.88 dBA

(C) 87.50 dBA

(D) 81.25 dBA

PROBLEM 3.9 *Crane Hands Signal*

The below hand signal by the signal person, with the arm extended horizontally to the side and the thumb pointing down with the other fingers closed, during a crane's operation means:

(A) Lower the hoist

(B) Bring load down

(C) Lower the boom

(D) Lower the boom and raise the load

PROBLEM 3.10 *Trench Support*

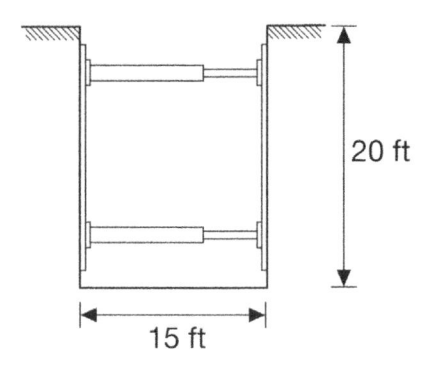

The above sketch is for a 20 ft deep trench excavated in soil type B. The deep trench is protected using plywood and aluminum hydraulic shoring as shown.

The horizontal interval this shoring should be placed at is most nearly:

(A) 8 ft

(B) 6.5 ft

(C) 5.5 ft

(D) 4 ft

PROBLEM 3.11 *Soil Loss Prevention*

The following bare land properties are provided for soil loss calculation and prevention:

- 300 ft long sloped at 2%
- Type of soil is Sandy Loam with 0.5% organic matter
- Rainfall intensity index is 200 $ton.in/acre.yr$

Given a permissible soil loss limit of 11 $tons/hectare.yr$, the conservation factor should be:

(A) 0.3

(B) 0.74

(C) 0.014

(D) 0.033

PROBLEM 3.12 *Soil Erodibility*

Which of the following contributes to the amount of soil loss caused by erosion:

I. Porosity: porosity affects soil structure. The more porous the soil, the weaker its structure becomes, leading to increased erodibility.

II. Reduction in vegetation cover contributes to erosion.

III. An increase in kinetic energy from both wind and rainfall leads to higher erosion rates.

IV. Runoff distance plays a role. Longer runoff routes (i.e., longer lands) result in reduced erosion.

V. Runoff slope. The steeper the land slope, the greater the expected erosion.

(A) I + II + III + IV + V

(B) II + III + V

(C) I + II + III + V

(D) II + III + IV + V

PROBLEM 3.13 *Excavation Slope*

The allowable slope angle for an excavation dug in type C soil is:

(A) 53°

(B) 45°

(C) 34°

(D) 63°

PROBLEM 3.14 *Construction Methods*

Choose three from the below list of construction machinery used in asphalting:

☐ Motor grader
☐ Paver screed
☐ Pneumatic tire roller
☐ Milling Machine
☐ Double steel roller
☐ Single smooth drum roller

PROBLEM 3.15 *Excavation for a Deep Pumping Station*

A deep pumping station is to be constructed prior to the preliminary works of a water treatment plant with the aim of lifting the sewage which is gravitated over a long distance by 75 ft. The pumping station is circular in shape with a diameter of 45 ft.

To construct this pumping station, deep excavation has to be performed in a medium dense soil. The site is not located in a populated area and hence there is nothing to account for around the pumping station location apart from providing a proper working space.

The most economical method to perform this excavation in preparation to the construction works for this pumping station is the following:

(A) Provide a secant pile wall at the permitter of the excavation, the secant pile depth is to be extended below the bottom of the excavation to control groundwater if any.

(B) Provide cantilevered sheet piles at the perimeter of the excavation, sheet piles can be used as formwork that the exterior wall of the pumping station can be poured against.

(C) Provide a deep diaphragm/slurry wall at the perimeter of the excavation to be extended below the excavation. The pumping station exterior walls can be cast against the diaphragm wall after excavation.

(D) The safest method, and given that there is no workspace limitation, is an open cut excavation, proper dewatering pumps can be provided to control groundwater if any.

PROBLEM 3.16 *Equipment Operations Near Power Lines*

According to Occupational Health and Safety Regulations, the minimum distance that must be maintained between a crane and a power line energized with 200 kV to 350 kV is:

(A) 10 ft

(B) 20 ft

(C) 30 ft

(D) 40 ft

PROBLEM 3.17 *Volume of Excavation*

The below cut and fill diagram belongs to a highway project. The project's existing ground consists of soil with 20% shrinkage factor and 15% swell factor.

The diagram's negative y-axis represents cross section areas to be cut and the positive y-axis represents cross areas sections to be filled.

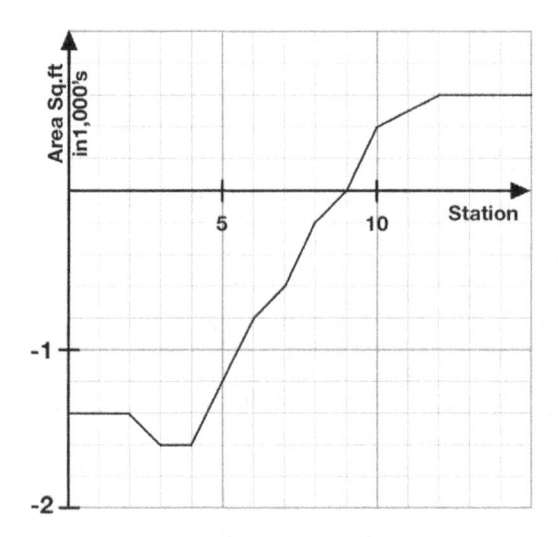

The project shall balance its cut and fill material, and the rest goes to waste.

Considering dump trucks can haul up to 11 $yard^3$ per trip, the number of trips expected to haul waste outside the project is most nearly:

(A) 2,470

(B) 2,190

(C) 1,240

(D) 2,680

PROBLEM 3.18 *Cut Volume*
Below is an excavation profile with the cross-sectional cut area in ft^2 noted at every interval. The distance between them is $25\ ft$.

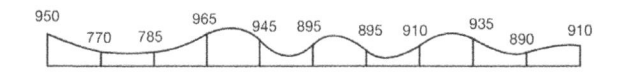

The total volume of cut for this profile is most nearly:

(A) $222,500\ ft^3$

(B) $667,500\ ft^3$

(C) $295,250\ ft^3$

(D) $332,355\ ft^3$

PROBLEM 3.19 *Borrow Pit Volume*
The below is a borrow pit. The major grid is $25\ ft \times 25\ ft$. The elevation from the ground level is mentioned on the grid as shown with the surrounding ground being level.

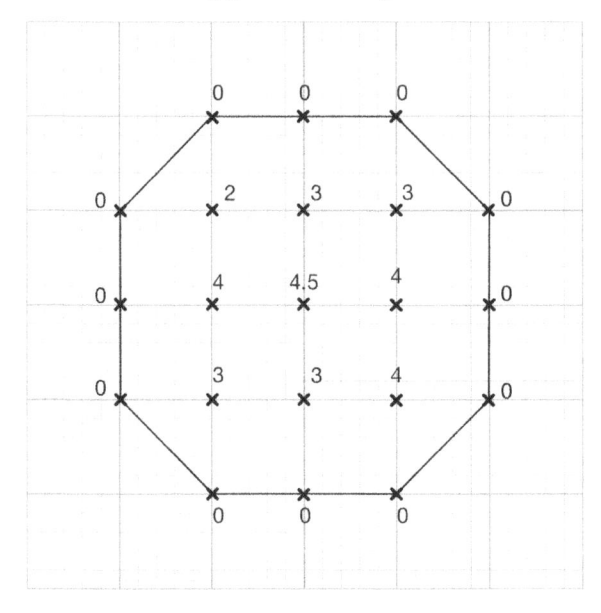

The volume of this borrow pit is most nearly:

(A) $18,500\ ft^3$

(B) $12,500\ ft^3$

(C) $32,500\ ft^3$

(D) $23,500\ ft^3$

PROBLEM 3.20 *Measuring Groundwater*
The below instrument can be used to measure negative water pressure in the ground:

(A) Open standpipe piezometer

(B) Pneumatic piezometer

(C) Vibrating wire piezometer

(D) Observation well

PROBLEM 3.21 *Water Measurement*
The below method or device is the most accurate method used to measure water level in observation wells:

(A) Chalked tape

(B) Tape with a float

(C) Electric water-level indicator

(D) Data loggers

PROBLEM 3.22 *Mass Haul Diagram*
The below Mass Haul Diagram belongs to a highway with soil shrinkage factor of 14.5%.

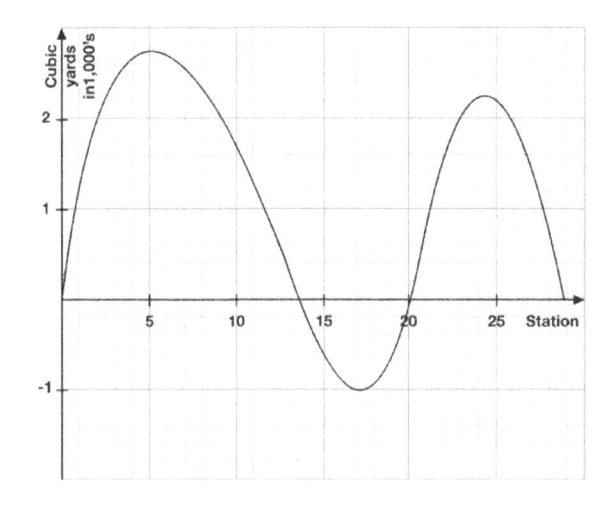

Considering a Free Haul Distance of 500 ft and a Free Haul Volume of 1,600 $yard^3$, wastage in $yard^3$ for this project is most nearly:

(A) 800

(B) 400

(C) 1,400

(D) 685

PROBLEM 3.23 *Scour Protection (1)*
In a river restoration project, where the riverbanks are prone to high flow velocities and sediment transport due to the zigzag pattern of the river, the best scour protection that can reduce or eliminate further banks scouring is the following:

(A) Planting native riparian vegetation along the riverbanks promoting root growth to stabilize the soil and reduce erosion.

(B) Installing rock vanes or groins perpendicular to the flow redirecting the flow away from the vulnerable riverbank areas to reduce its erosion.

(C) Construct gabion baskets filled with rocks along the riverbanks providing flexible erosion protection and stabilizing sediments.

(D) Implementing geotextile fabric layer anchored with vegetation reinforcing the riverbank and preventing soil loss.

PROBLEM 3.24 *Scour Protection (2)*
A bridge that passes over a river has one of its piers located in a region of the river known for its erodible soils and frequent floods.

Based on this information, the following scour protection measure is the most appropriate to ensure long-term stability for the bridge and its pier:

(A) Riprap

(B) Concrete aprons

(C) Sacrificial piles

(D) Underwater concrete

PROBLEM 3.25 *Ground Measurements*
A deep excavation that is being dug adjacent to an existing building requires careful monitoring.

The most appropriate instrument for tracking lateral movements of the building during and after the excavation is:

(A) Settlement plate

(B) Inclinometer

(C) Crack gauge

(D) Tiltmeter

SOLUTION 3.1

The *NCEES Handbook,* Chapter 3 Geotechnical, Section 3.9 Laboratory and Field Compaction is referred to.

Optimum moisture content occurs at the soil's maximum dry density γ_d. In which case, dry density is calculated using the total (moist) density γ_t and moisture content w as follows:

$$\gamma_d = \frac{\gamma_t}{(1+w)}$$

Density is derived from the volume of the mold provided in the *Handbook* as $1/30 \, ft^3$, based upon which, compaction curve is created as shown below:

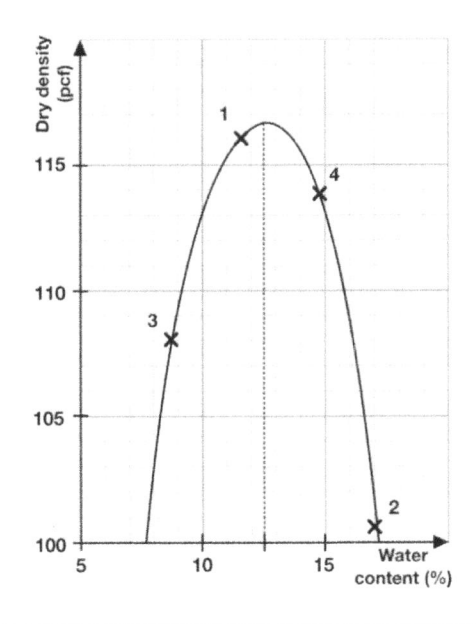

SN	wt_{moist} lb	γ_t lb/ft³	w %	γ_d lb/ft³
1	4.32	129.60	11.6%	116.13
2	3.92	117.60	17.0%	100.51
3	3.92	117.60	8.8%	108.09
4	4.36	130.80	14.8%	113.94

The optimum moisture content as derived from the curve is when dry density is max at 12.5%.

Correct Answer is (C)

SOLUTION 3.2

This question uses the *CFR Title 29 2020, Part 1926 Safety and Health Regulations for Construction.* This document can be downloaded or can be viewed online and knowledge of it is required for this exam.

The provided cohesive soil, with an unconfined compressive strength of 45 kPa, corresponds to **Soil Type C** as defined in Subpart P, Appendix A (page 351).

Appendix C of Subpart P outlines the timber shoring arrangements for trenches. Specifically, Table C1.3 in this section details the minimum timber requirements for Soil Type C.

According to this table, the following arrangement is most suitable for trenches that are 20 ft deep and 15 ft wide:

10×10 cross braces with 6 ft horizontal spacing and 5 ft vertical spacing.

Correct Answer is (A)

SOLUTION 3.3

Equations from the *NCEES Handbook version 2.0,* Section 3.8.3 can be used to solve this question.

First let's compute the required void ratio at the embankment site e_E and compare it to the given void ratio at the borrow pit site $e_B = 0.85$

$$\gamma_{d,E} = \frac{G_s \gamma_w}{1 + e_E}$$

$$\rightarrow e_E = \frac{G_s \gamma_w}{\gamma_{d,E}} - 1$$

$$= \frac{2.7 \times 62.4 \, pcf}{100 \, pcf} - 1$$

$$= 0.68$$

Calculate the total volume in both locations using the volume of solids V_S given that the volume of solids between borrowing and dumping will not change, it is only the volume of voids that will change:

$$V_E = V_S(1 + e_E)$$

$$V_B = V_S(1 + e_B)$$

$$\rightarrow \frac{V_B}{V_E} = \frac{1 + e_B}{1 + e_E}$$

$$\frac{V_B}{200{,}000} = \frac{1 + 0.85}{1 + 0.68}$$

$$V_B = 220{,}238 \ ft^3 \ (8{,}157 \ yd^3)$$

$$Cost = 8{,}157 \times 2 = \$16{,}314$$

Correct Answer is (C)

SOLUTION 3.4
Equations from the *NCEES Handbook version 2.0* Section 3.8.3 can be used to solve this question.

The dry density of the borrow pit material is calculated as follows:

$$\gamma_{d,B} = \frac{\gamma_{t,B}}{1 + w}$$

$$= \frac{110}{1 + 8\%}$$

$$= 101.85 \ lb/ft^3$$

In which case, the weight of the solids for each $1 \ ft^3$ in the borrow area is:

$$W_{S,B} = 101.85 \ lb$$

As for the embankment:

$$w = 14\%$$

The additional quantity of water required is:

$$w_{add} = 14\% - 8\% = 6\%$$

Weight of water to be added in relation to the weight of solids for each $1 \ ft^3$ in the borrow area is:

$$W_w = w_{add} W_{S,B}$$

$$= 6\% \times 101.85 \ lb$$

$$= 6.11 \ lb \ in \ ft^3 \ (165 \ lb \ in \ yd^3)$$

Convert $165 \ lb$ of water into volume using water density of $62.4 \ lb/ft^3$ → $2.64 \ ft^3$ ($19.75 \ gals$).

Correct Answer is (B)

SOLUTION 3.5
Equations from the *NCEES Handbook version 2.0* Section 5.5.6.1 can be used to solve this question.

The effective specific gravity of aggregate is calculated as follows:

$$G_{se} = \frac{100 - P_b}{\frac{100}{G_{mm}} - \frac{P_b}{G_b}}$$

P_b is the asphalt percentage in the mix, which makes $(100 - P_b)$ the percentage of stones.

G_{mm} is the max specific gravity of the paving material with no air voids.

$$= \frac{95.2}{\frac{100}{2.55} - \frac{4.8}{1.05}}$$

$$= 2.75$$

Correct Answer is (D)

SOLUTION 3.6
Safety Incidence Rate *IR* equation can be picked up from the *NCEES Handbook* Section 2.6.1.1.

$$IR = \frac{N \times 200{,}000}{T}$$

$$IR_{1st\ project} = \frac{250 \times 200{,}000}{0.85 \times 2 \times 12 \times 260 \times 150}$$

$$= 62.85$$

$$N_{2nd\ project} = \frac{IR_{first\ project,improved} \times T}{200{,}000}$$

$$= \frac{(0.75 \times 62.85) \times 0.85 \times 1.5 \times 12 \times 260 \times 75}{200{,}000}$$

$$= 70$$

Correct Answer is (B)

SOLUTION 3.7
The *NCEES Handbook*, Chapter 3 Geotechnical, Section 3.10 Trench and Excavation Construction Safety is referred to.

Soil type C should be excavated with a slope of 1 vertical to 1 ½ horizontal as shown below:

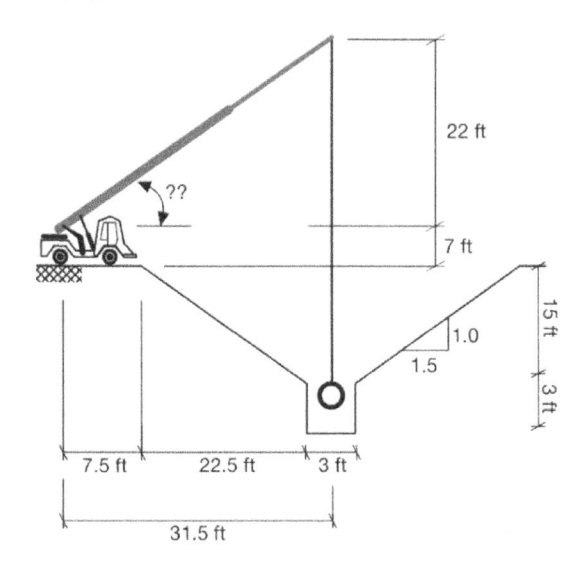

With the above slope configuration, all missing dimensions can be determined, and the required angle can be calculated using trigonometry as follows:

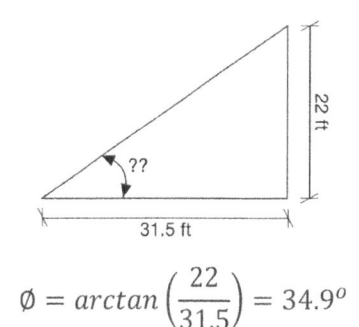

$$\emptyset = \arctan\left(\frac{22}{31.5}\right) = 34.9°$$

Correct Answer is (B)

SOLUTION 3.8
The *NCEES Handbook,* Chapter 2 Construction, Section 2.6.2 Work Zone and Public Safety is referred to. The following equation along with the permissible timetable are referred to:

$$D = 100 \times \sum \frac{C_i}{T_i}$$

C_i is time spent at the specified noise pressure. T_i is permissible time taken from the OSHA's table for each exposure specified above represented by i.

$$D = 100 \times \left(\frac{2}{32} + \frac{2}{16} + \frac{3}{8} + \frac{1}{4}\right) = 81.25\ dBA$$

Correct Answer is (D)

SOLUTION 3.9
The *CFR Title 29 2020, Part 1926 Safety and Health Regulations for Construction* is used in this question. This document can be downloaded online and knowledge of it is required for this exam.

Subpart CC Appendix A of page 742 defines the standard hand signals for a crane operation, in which case, an extended hand with the thumb pointing down means **"lower the boom"**.

Correct Answer is (C)

SOLUTION 3.10

The *CFR Title 29 2020, Part 1926 Safety and Health Regulations for Construction* is used in this question. This document can be downloaded online and knowledge of it is required for this exam.

Subpart P Appendix D Aluminum Hydraulic Shoring for Trenches, the relevant table, Table D1.2 of page 372 states the horizontal placement for soil type B, 20 ft deep, should be every 5.5 ft.

Correct Answer is (C)

SOLUTION 3.11

The *NCEES Handbook version 2.0*, Chapter 6 Water Resources and Environmental, Section 6.5.9.2 Erosion/ Revised Universal Soil Loss equation of page 411 is referred to.

$$A = R.K.LS.C.P$$

P is the conservation factor, given all inputs of this equation are provided in the body of the question, P is calculated as follows:

$$P = \frac{A}{R.K.LS.C}$$

A is the amount of soil loss due to erosion measured in *tons per acre per year*:

$$A = 11 \frac{tons}{hectare.yr} \times \frac{1\ hectare}{2.47\ acre}$$
$$= 4.45 \frac{tons}{acre.yr}$$

K is the soil erodibility factor taken from the same section of *NCEES Handbook version 2.0* page 412 as '0.27' for Sandy Loam with 0.5% organic matter.

LS is the topographic factor taken from the same section of *NCEES Handbook version 2.0* page 412 as '0.28' for a 300 ft land sloped at 2%.

C is the crop and cover management factor taken as '1.0' for bare land.

$$P = \frac{A}{R.K.LS.C}$$
$$= \frac{4.45}{200 \times 0.27 \times 0.28 \times 1.0}$$
$$= 0.29\ (*)$$

Correct Answer is (A)

(*) A conservation value of $P = 0.29$ requires strip cropping and contour farming. In a nutshell this requires growing crops in a systematic arrangement of strips along the contours and across a sloping field.

SOLUTION 3.12

The *NCEES Handbook version 2.0*, Chapter 6, Section 6.5.9.2 Erosion/ Revised Universal Soil Loss equation is referred to in order to provide context into the solution.

The Revised Universal Soil Loss equation:

$$A = R.K.LS.C.P$$

A is the amount of soil loss due to erosion (*tons per acre per year*), R is the rainfall erosion index or the climatic erosivity, K is soil erodibility factor, LS is the topographic factor and is taken from the table provided in page 412 of the handbook's version 2.0. C represents vegetation and is called the crop and cover management factor, and P is the erosion control practices factor.

Factors Influencing Soil Loss:

Rainfall erosion index R considers the intensity, duration, and continuity of rainfall. Rainfall erosivity and its relationship to kinetic energy play a crucial role in erosion. **This marks statement III as correct.**

Soil erodibility K is its susceptibility to erosion. Factors affecting K include soil aggregation and structure. The more porous the soil, the reduced runoff it shall experience and the lesser effect it would have on its continuous erodibility.
This marks statement I as incorrect.

In a similar fashion, vegetation cover acts as a barrier against erosion by obstructing water velocity, hence lesser runoff is experienced with more cover.
This marks statement II as correct.

Topographic factor LS, specifically the slope length and its steepness, are both proportional to erosion. Which means, more length and more slope causes more erosion.
This marks statement IV as incorrect, however, it makes Statement V true.

Correct Answer is (B)

SOLUTION 3.13
The *NCEES Handbook version 2.0* Section 3.10.2 is referred to in which case a slope for an excavation in types C soil is defined as $34°$.

Correct Answer is (C)

SOLUTION 3.14
This question requires more of a hands-on field experience and is not a typical question that can be found in a handbook.

Construction machinery used for the purpose of asphalting are as follows:

- ☑ **(Double) Steel wheel roller**
  Steel wheel rollers are self-propelled compaction devices that use steel drums to compress hot asphalt mixes. Usually one, two or three drums. Two drums are most used.

- ☑ **Paver screed**
  Paver screed receives the hot mix from dumping trucks, distributes it and paves it to the required thickness. Pavers have enough power to push the truck as it empties its content into the paver's receiving shaft. Pavers control the speed by which the truck moves, and it can achieve 70-80% density of the layer.

- ☑ **Pneumatic tire roller**
  Those are self-propelled compaction devices that have pneumatic tires which provide proper and smooth compaction to the layer beneath as those tires have no threads in them.

The following machinery are not commonly used in a normal asphalting operation:

- ☐ **Milling machine**
  Milling machines are used to cut existing asphalt or remove a top layer of the existing pavement to create a bed surface. In this case it cannot be claimed to be useful during an asphalting operation unless specifically required on the field.

- ☐ **Single smooth drum roller**
  Smooth drum rollers use static and vibratory pressure to compact rough material such as gravel, rocks, and sand. They are more effective in granular material, and they would not normally produce a smooth surface. A single drum roller has tires at the back and those tires come with deep threads which are not suitable for finishing asphalt layers.

- ☐ **Motor grader**
  Motor graders have an adjustable blade that can be used to complete various

construction activities such as surface leveling, fine grading, creating slopes, creating ditches and earth moving. Graders have tires with deep threads which makes it challenging to work on top of a hot mix asphalt. Graders can however be used for asphalting during abnormal circumstances, such as during a breakdown of the asphalt paver, or paving steep slopes, or locations with narrow access that screeds cannot enter

SOLUTION 3.15

A secant pile wall (*) is constructed by interlocking vertical piles, forming a barrier, or retaining structure. These walls find application in deep excavations. However, beyond a depth of $50\,ft$ to $55\,ft$, their feasibility diminishes. Not only that, but their effectiveness also decreases, and the risk of failure increases significantly.

Another option to perform deep excavations is the cantilevered sheet pile wall. However, these sheet piles are effective up to a depth of $10\,ft$ to $15\,ft$. Beyond this depth, without proper bracing (which is not suitable for a $45\,ft$ wide excavation in this case), they fail to perform as intended and may/will collapse.

Diaphragm walls, on the other hand, are created using a slurry machine. These thin and very deep walls utilize bentonite during excavation. The bentonite adheres to the earth-dug wall sides, preventing collapse before casting. By adding reinforcements in the form of a cage prior to concreting, the wall's strength, rigidity, and depth are improved, making it suitable for very deep applications. At such depths, diaphragm walls can serve as sacrificial walls against which the pump wall can be cast if needed. Additionally, applying a waterproofing membrane to the diaphragm wall's internal face effectively protects the exterior face for the pump walls.

An open excavation can be deemed one of the safest and most traditional methods when all safety protocols are adhered to. However, it requires significant space, and with this depth, significant volume of excavation will be required. For instance, if the soil type necessitates a 45 degree cut, a pumping station this deep adds an extra excavation of $850k\,ft^3$ when the volume of excavation for the pump alone (if using the diaphragm wall method) is nearly $120k\,ft^3$.

Based on the above explanation, **the diaphragm wall method can be considered one of the economical methods for this application considering the amount of excavation that will be saved.**

Correct Answer is (C)

(*) Several references can be used to in order to expand knowledge in this area and other similar areas such as the UFC 3-220-10 Table 6-9 page 313, and the UFC 3-220-05 Section 2-10.

SOLUTION 3.16

The *CFR Title 29 2020, Part 1926 Safety and Health Regulations for Construction* is used in this question. This document can be downloaded online and knowledge of it is required for this exam.

Section 1926.1408 Clause (a)(2)(ii) option 2 of page 688, also Table A – Minimum Clearance Distance of page 690 specify the clearance distance for equipment operating near an energized line with power $200\,kV$ to $350\,kV$ as $20\,ft$.

Correct Answer is (B)

SOLUTION 3.17

The average area method is used to measure the volume for each two consecutive cross-sectional areas where the distance between consecutive areas is a station = 100 ft:

$$V = L\left(\frac{A_1 + A_2}{2}\right)$$

The following two tables represent cross sections measured at each station along with the volume between every two consecutive stations.

Bank (undisturbed) cut volume $V_{B,cut} = -950,000 \, ft^3$ is converted to lose volume $V_{L,cut}$. Fill requirements $V_{B,fill}$ is also converted into loose volume $V_{L,fill}$. The balance of both goes to waste.

Station	Cut Area	Fill Area
	ft^2	ft^2
0+00	− 1,400	
1+00	− 1,400	
2+00	− 1,400	
3+00	− 1,600	
4+00	− 1,600	
5+00	−1,200	
6+00	− 800	
7+00	− 600	
8+00	− 200	
9+00	0	0
10+00		400
11+00		500
12+00		600
13+00		600
14+00		600
15+00		600

Station	Cut Volume	Fill Volume
	ft^3	ft^3
0+00 to 1+00	− 140,000	
1+00 to 2+00	− 140,000	
2+00 to 3+00	− 150,000	
3+00 to 4+00	− 160,000	
4+00 to 5+00	− 140,000	
5+00 to 6+00	−100,000	
6+00 to 7+00	− 70,000	
7+00 to 8+00	− 40,000	
8+00 to 9+00	− 10,000	0
9+00 to 10+00	0	20,000
10+00 to 11+00		45,000
11+00 to 12+00		55.000
12+00 to 13+00		60,000
13+00 to 14+00		60,000
14+00 to 15+00		60,000
Total	− 950,000	300,000

$$V_{L,cut} = \left(1 + \frac{S_w}{100}\right) V_{B,cut}$$

$$= \left(1 + \frac{15}{100}\right)(-950,000)$$

$$= -1,092,500 \, ft^3$$

$$V_{L,fill} = \left(1 + \frac{S_h}{100}\right) V_{B,fill}$$

$$= \left(1 + \frac{20}{100}\right)(300,000)$$

$$= 360,000 \, ft^3$$

$$V_{L,waste} = -1,092,500 + 360,000$$

$$= -732,500 \, ft^3$$

$$No.\,of\,trips = \frac{732,500 \, ft^3}{11 \, yd^3 \times 27 \frac{ft^3}{yd^3}}$$

$$= 2,466 \, trip$$

Correct Answer is (A)

SOLUTION 3.18

The Simpson's rule which is presented in the *NCEES Handbook version 2.0* Section 2.1.2.3 can be used in this case especially that there are even number of intervals.

$$A = \frac{1}{3}\begin{bmatrix} 950 + 910 \\ + \\ 2 \times (785 + 945 + 895 + 935) \\ + \\ 4 \times (770 + 965 + 895 + 910 + 890) \end{bmatrix} \times 25$$

$$= 222{,}500 \, ft^3$$

Correct Answer is (A)

SOLUTION 3.19

The grid method which is presented in the *NCEES Handbook version 2.0* Section 2.1.2.2 can be referred to in this question.

The below grid is numbered per the following sketch where:

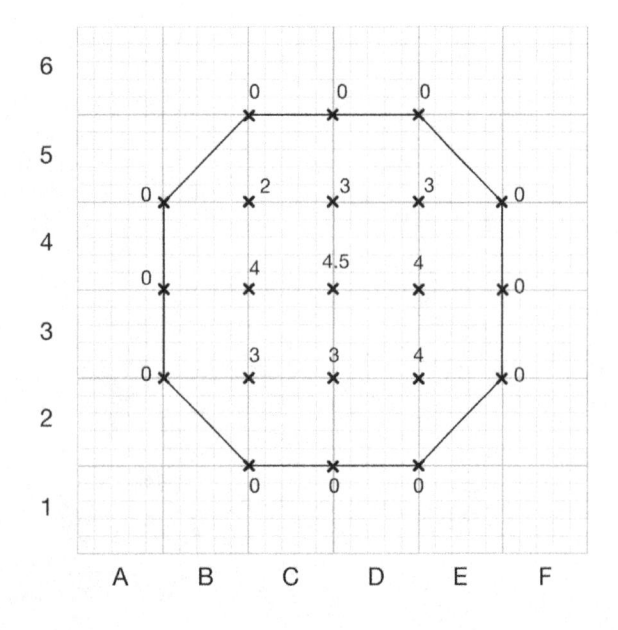

- C3, C4, D3 & D4 are middle grids.
- C2, D2, C5, D5, B3, B4, E3 & E4 are edge wedges.
- B2, E2, B5 & E5 are corner wedges.

Calculate the volume of the four middle girds starting from the top left and moving clockwise:

$$V = \begin{bmatrix} \frac{1}{4} \times (2 + 3 + 4 + 4.5) \times 25 \times 25 \\ + \\ \frac{1}{4} \times (3 + 3 + 4.5 + 4) \times 25 \times 25 \\ + \\ \frac{1}{4} \times (4 + 4.5 + 4 + 3) \times 25 \times 25 \\ + \\ \frac{1}{4} \times (4.5 + 4 + 3 + 3) \times 25 \times 25 \end{bmatrix}$$

$$= 9{,}062.5 \, ft^3$$

An edge wedge is sketched below to the left while a corner wedge is sketched to the right:

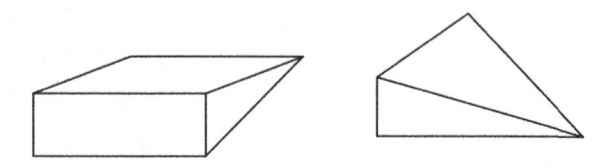

Calculate the volume of the eight edge wedges starting from the top left and moving clockwise:

$$V = \begin{bmatrix} \dfrac{(0.5 \times (2+3) \times 25 \times 25)}{2} \\ + \\ \dfrac{(0.5 \times (3+3) \times 25 \times 25)}{2} \\ + \\ \dfrac{(0.5 \times (3+4) \times 25 \times 25)}{2} \\ + \\ \dfrac{(0.5 \times (4+4) \times 25 \times 25)}{2} \\ + \\ \dfrac{(0.5 \times (3+4) \times 25 \times 25)}{2} \\ + \\ \dfrac{(0.5 \times (3+3) \times 25 \times 25)}{2} \\ + \\ \dfrac{(0.5 \times (3+4) \times 25 \times 25)}{2} \\ + \\ \dfrac{(0.5 \times (4+2) \times 25 \times 25)}{2} \end{bmatrix}$$

$$= 8{,}125.0 \, ft^3$$

Calculate the volume of the four corner wedges starting from the top left and moving clockwise:

$$V = \begin{bmatrix} \dfrac{0.5 \times 2 \times 25 \times 25}{3} \\ + \\ \dfrac{0.5 \times 3 \times 25 \times 25}{3} \\ + \\ \dfrac{0.5 \times 4 \times 25 \times 25}{3} \\ + \\ \dfrac{0.5 \times 3 \times 25 \times 25}{3} \end{bmatrix}$$

$$= 1{,}250.0 \, ft^3$$

The total volume of this borrow pit is:

$$V_{total} = 9{,}062.5 + 8{,}125.0 + 1{,}250.0$$

$$= 18{,}437.5 \, ft^3$$

Correct Answer is (A)

SOLUTION 3.20

Reference is made to the *FHWA NHI-16-072 Geotechnical Site Characterization*, Section 10.6 Investigation of Groundwater Conditions using Piezometers.

Table 10-2 of this section defines piezometer types and their applications. The **vibrating wire piezometer** in this case is the one used to measure negative water pressure.

Correct Answer is (C)

SOLUTION 3.21

Reference is made to the *FHWA NHI-06-088 Soils and Foundations Reference Manual Volume I,* Section 3.13.4 Water Level Measurements.

Reading through the above referenced section, the **chalked tape**, albeit the cheapest method, is the most accurate one.

Correct Answer is (A)

SOLUTION 3.22

The Free Haul Distance (FHD) has been given in the problem as $500\ ft$. The FHD is the distance below which earthmoving is considered part of the contract and contractor cannot claim for extras for overhauling.

To identify stations which fall within the FHD, a $500\ ft$ *to-scale* horizonal line is drawn and fit in position to intersect close to the peaks and troughs of the Mass Haul Diagram (MHD) curves as shown in the figure below. The y-axis generated values from the FHD intersection with the MHD curves represent the quantity which will be hauled as part of the contract price with no extras.

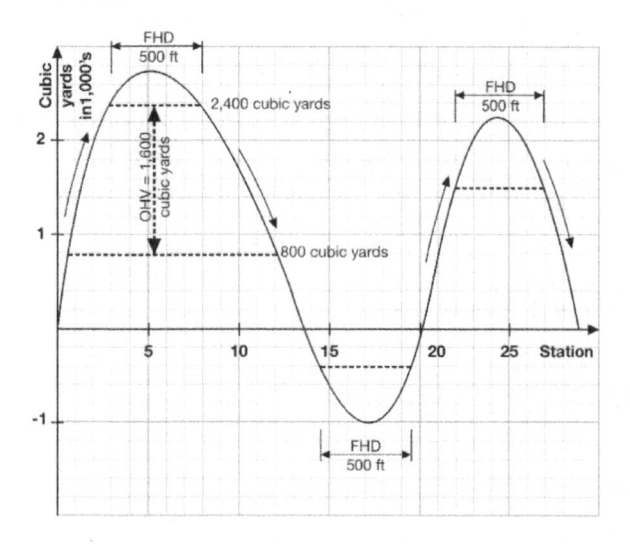

The Over Haul Volume (OVH) is the volume beyond which earthmoving can be claimed as extra by the contractor. OVH is the vertical distance from the FHD towards the x-axis (whether upwards or downwards) to an imaginary horizontal line that intersects with the two sides of the semi parabolic curves.

In this case, and as given in the question, the $OHV = 1,600\ yard^3$ measured as a vertical distance from the FHD down to an intersection of $800\ yard^3$ for the first MHD curve. This does not apply to the second or the third curves because there isn't enough vertical distance to establish such ordinates.

The vertical ordinates generated form this intersection represent either: (1) the waste material when the curve is moving upwards (i.e., cut sections which is more economical to dump outside the project), or (2) the borrow material when the curve is moving downwards (i.e., fill sections which is more economical to supply its material from outside the project). The horizonal intersection is hence called the Limit of Economic Hall (LEH). Shrinkage does not apply in the cut section therefore, **wastage in this case will only be $800\ yard^3$** (*).

Correct Answer is (A)

(*) To put things into perspective for this question, and as concluded from the MHD, the material that should go to waste is located between stations $'0 + 00'$ and $'0 + 70'$. If we assume this is a $60\ ft$ wide highway construction project, the height of this cut section that should go to waste is nearly:

$$\frac{800\ yard^3 \times 27\frac{ft^3}{yard^3}}{60\ ft\ \times 70\ ft} \approx 5\ feet$$

SOLUTION 3.23

The best option in this case is option (B), which involves installing rock vanes.

Rock vanes, or groins, are structures placed almost perpendicular to the flow of water and they are effective in redirecting flow away from vulnerable riverbanks.

Vanes can force flow away from riverbanks, reducing flow velocities near the sides and increasing them in the center.

The sketch below provides a high-level explanation of how these vanes look like:

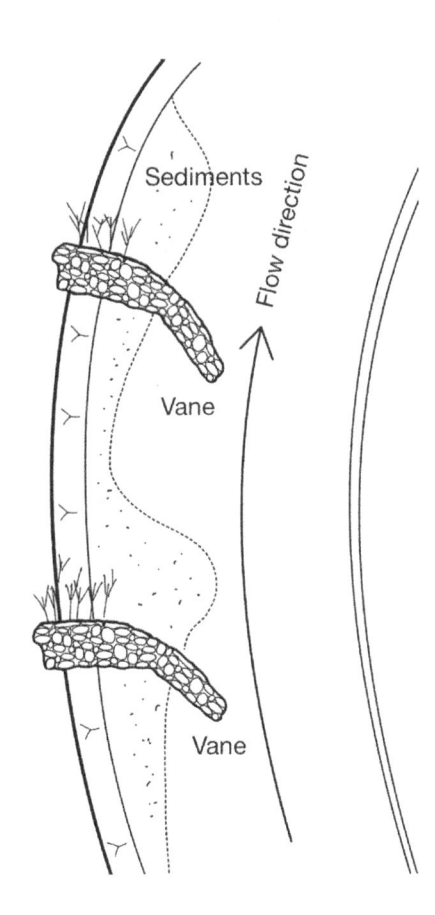

On the other hand, planting vegetation can help stabilize the soil, but it does not effectively address high flow velocities as over time, vegetation can be washed away. Gabion baskets or geotextiles can provide erosion protection, but rock vanes are more effective as they can alter high velocities into lower velocities, and aesthetically, they look more integrated with the environment.

You may find other scour protection measures in the references provided for the exam. However, these measures are limited, and during the exam, you may encounter questions not covered in these references, especially on this subject. Therefore, it is advised to conduct further research on scour protection to understand the available measures and their applications (*).

Correct Answer is (B)

(*) Below are some measures used for scour protection for bridge piers, culverts and riverbanks along with some of their applications:

Riprap for Bridge Piers:
Large stones placed around bridge piers to absorb and deflect water energy, preventing erosion directly at the pier base.

Gabions for Riverbanks:
Wire mesh baskets filled with rocks, placed alongside riverbanks to stabilize and protect against erosion caused by flowing water.

Geotextiles for Culverts:
Synthetic fabrics used to line the soil around culvert inlets and outlets, preventing erosion and soil displacement.

Concrete Aprons for Culverts:
Concrete slabs installed at culvert inlets and outlets to provide a hard, erosion-resistant surface, protecting against high-velocity flows.

Rock Vanes for Riverbanks:
Structures placed perpendicular to the flow to redirect water away from eroding riverbanks, reducing flow velocities near the banks.

Vegetation for Riverbanks:
Planting deep-rooted vegetation along riverbanks to stabilize soil and reduce surface runoff, effective in low to moderate flow conditions.

Sheet Piles for Bridge Piers:
Vertical barriers driven into the riverbed around bridge piers to prevent soil movement and provide structural support against scour.

Articulated Concrete Blocks for Culverts:
Interlocking concrete blocks forming a flexible mat, placed at culvert outlets to protect against erosion from high-velocity discharges.

Bendway Weirs for Riverbanks:
Low-profile structures angled to the flow, redirecting water away from eroding banks and stabilizing the river channel.

Spur Dikes for Bridge Piers:
Structures extending from the riverbank into the river to redirect flow away from bridge piers, reducing local scour around the piers.

SOLUTION 3.24

Reference could be made to some of the measures presented and discussed in Solution 3.23, along with the new measures introduced in this question. Among these, **riprap** appears to be the most applicable solution for this case.

Riprap typically consists of materials such as loose rock, concrete blocks, or other durable substances designed to absorb and deflect the energy of flowing water.

However, certain design considerations must be considered. In some circumstances, placing a riprap blanket around a bridge pier can promote erosion around the edges of the armored zone. This can cause the riprap covering to fail if the channel deepens sufficiently in the unprotected areas between piers. Therefore, special attention should be given to the size, angle of placement, and velocity of flow.

See below a representative sketch for the riprap around a pier along with some elements that shall be considered during design:

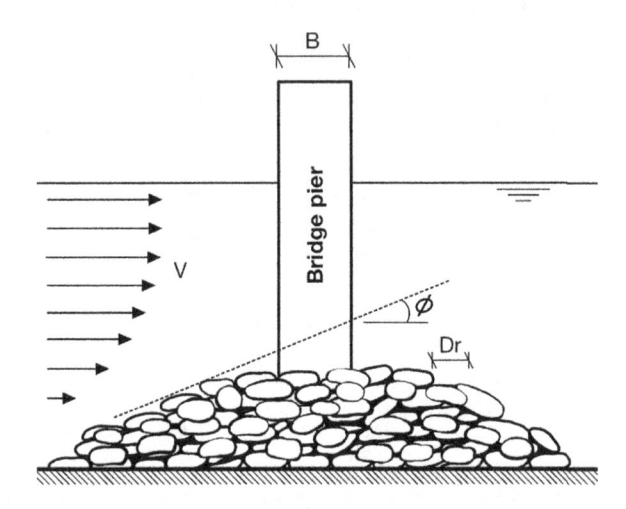

On the other hand, concrete aprons are more suitable for culverts and similar applications. Sacrificial piles, which are additional piles designed to absorb the impact of scour, may not be as effective as riprap in this scenario, however, they are effective in stabilizing soil and preventing its movement. Underwater concrete can be used to stabilize the riverbed and protect against erosion, but it is not applicable in this case and it does not provide the required long-term support.

Correct Answer is (A)

SOLUTION 3.25

Inclinometers are used to measure horizontal or rotational movements, making them suitable for monitoring lateral movements of an existing building's foundation caused by nearby activities.

Settlement plates, on the other hand, measure vertical settlements, while crack gauges measure crack widths, and tiltmeters measure small changes in tilt or rotation.

Correct Answer is (B)

PART III
Const. Observation

IV
EARTHQUAKE ENGINEERING

Knowledge Areas Covered

SN	Knowledge Area
4	**Earthquake Engineering and Dynamic Loads** A. Seismic site characterization B. Seismic analyses and design (e.g., liquefaction, pseudo static, earthquake loads)

PART IV
Earthquake Eng.

PROBLEM 4.1 *Physical Properties Depth*

A soil profile is to be developed for seismic analysis for a certain location where the bedrock is quite deep and unreachable.

The depth over which the soil physical properties for seismic analyses should be analyzed in this case is at least:

(A) 50 ft

(B) 100 ft

(C) 150 ft

(D) 200 ft

PROBLEM 4.2 *Shear Wave Velocity Test*

The below method provides accurate results when measuring a shear, or compressional, wave velocity for a certain location:

(A) Cross-hole survey using the borehole method.

(B) Down-hole survey using the borehole method.

(C) Up-hole survey using the borehole method.

(D) Seismic refraction method.

PROBLEM 4.3 *Initial Shear Modulus*

A Standard Penetration Test (SPT) was performed at a certain depth for one of the locations where it is required to measure the initial shear modulus G_{max} at. The standardized blow count at the location was $N_{60} = 15$, and the Placidity Index $PI = 0$.

Based on this information, the initial shear modulus G_{max} is most nearly:

(A) 100,000 kPa

(B) 150,000 kPa

(C) 200,000 kPa

(D) 250,000 kPa

PROBLEM 4.4 *Shallow Foundation*

The following 6 ft × 6 ft square footing is constructed on soil and is loaded as follows:

$$V = 80 \; kip$$

$$M_x = 40 \; kip.ft \text{ due to seismic loading}$$

$$M_y = 20 \; kip.ft \text{ due to seismic loading}$$

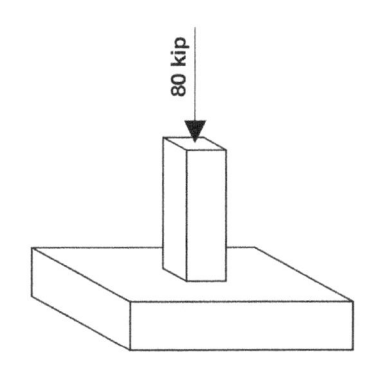

The equivalent uniform load below this foundation due to the above seismic effect is most nearly:

(A) 1.0 ksf

(B) 2.0 ksf

(C) 3.0 ksf

(D) 4.0 ksf

PROBLEM 4.5 *Adjusted Base Shear*

A five-story building constructed on a soil type C has a base shear of 400 kip. The building is made of steel and concrete composite ordinary braced frame.

Considering the effects of Soil Structure Interactions SSI, the lower limit that base shear cannot be reduced below is:

(A) 280 kip

(B) 320 kip

(C) 340 kip

(D) 360 kip

PROBLEM 4.6 *Site Classification*

The below is a borehole log for a borehole taken at a certain location where it is required to classify the site to carry out seismic analysis and design.

BOREHOLE LOG SHEET			
Depth (ft) / Groundwater Level / Graphic Log		Material Description	Samples Test Remarks
0–30		SANDY SILTY CLAY, medium plasticity, well graded with traces of poorly graded gravel. Density = 112 pcf	SPT N = 8 blows per ft
30–70	Water level at 4 ft	GRAVELLY SANDY CLAY, medium plasticity, motted brown, moist to wet and fully submerged at water level at 4 ft depth. Density = 118 pcf	SPT N = 30 blows per ft
70–100		VERY DENSE SOIL, Low plasticity, poorly graded with traces of well graded gravel, cohesionless Density = 120 pcf	SPT N = 55 blows per ft

Based on the above information, the site classification for this location is:

(A) Site C – very dense soil and soft rock

(B) Site D – Stiff soil

(C) Site E – Soft clay soil

(D) Site F – Vulnerable to failure

PROBLEM 4.7 *Seismic Design Category*

A six-story retail building located in Albuquerque, New Mexico (*) is founded on soil type C and is built using steel and concrete composite ordinary braced frames.

The Seismic Design Category (SDC) for this building is:

(A) SDC A

(B) SDC B

(C) SDC C

(D) SDC D

(*) Use the following spectral parameters for Albuquerque, New Mexico:

$S_S = 0.43$

$S_1 = 0.12$

PROBLEM 4.8 *Liquefaction Sensitive Soil*

Which of the below soils is considered more sensitive to liquefaction when they have high water content:

(A) Soil A:
Liquid limit = 0.3
Liquidity index = 0.7
$(N_1)_{60} = 4$

(B) Soil B:
Liquid limit = 0.5
Liquidity index = 0.2
$(N_1)_{60} = 9$

(C) Soil C:
Liquid limit = 0.1
Liquidity index = 0.1
$(N_1)_{60} = 3$

(D) Soil D:
Liquid limit = 0.6
Liquidity index = 0.6
$(N_1)_{60} = 15$

(✱) **PROBLEM 4.9** *Liquefaction Potential*
The liquefaction potential is to be assessed at the bottom of the below sand layer with fines content < 5% as shown, knowing that this layer supports a foundation that weighs 1 ksf.

The groundwater level is 5 ft below this foundation. The density of sand and clay layers are both 120 pcf. The modeled earthquake magnitude is estimated as $M_w = 7.5$ with a Peak Ground Acceleration PGA of $a_{max} = 0.3g$ and a corrected shear wave velocity for this layer of $v_{s1} = 150\ m/sec$.

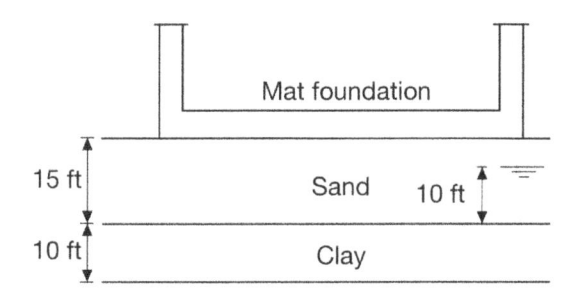

```
            Mat foundation
15 ft    Sand    10 ft
10 ft    Clay
```

Based on the above information, the value of the Cyclic Stress Resistance factor CSR and the potential of liquefaction for this soil layer is most nearly:

(A) 0.29 with liquefaction potential

(B) 0.29 with no liquefaction potential

(C) 0.58 with liquefaction potential

(D) 0.58 with no liquefaction potential

(✱) **PROBLEM 4.10** *Slope Stability Pseudo Static Seismic Coefficient*
To assess seismic slope stability, seismic coefficient k_s is used to determine the seismic load in the Capacity to Demand C/D ratio used in failure surface calculations.

Drawing from the above, a 65 ft tall slope located in Albuquerque, New Mexico, has the following seismic information:

o Site class D
o Peak Ground Acceleration $PGA = 0.3$

o Spectral acceleration at one second $S_1 = 0.2$

The seismic coefficient k_s for a required C/D ratio of 1.1 assuming 1 to 2 in of permanent seismic displacement for this slope is most nearly:

(A) 0.15

(B) 0.35

(C) 0.55

(D) 0.85

PROBLEM 4.11 *Av. Shear Wave Velocity*

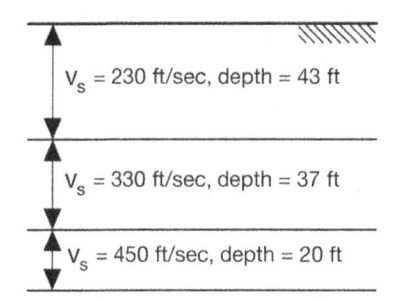

```
V_s = 230 ft/sec, depth = 43 ft
V_s = 330 ft/sec, depth = 37 ft
V_s = 450 ft/sec, depth = 20 ft
```

The average shear wave velocity $v_{s,av.}$ for the above soil layers is most nearly:

(A) 335 ft/sec

(B) 310 ft/sec

(C) 290 ft/sec

(D) 250 ft/sec

PROBLEM 4.12 *Shear Wave Velocity*
The maximum Young's modulus for a certain soil was measured using a cyclic soil test as $E_{max} = 2,500\ psi$, the soil's Poisson's ratio is 0.35, and its unit weight is 120 pcf.

Based on this information, the shear wave velocity v_s for this soil is most nearly:

(A) 20 ft/sec

(B) 190 ft/sec

(C) 250 ft/sec

(D) 310 ft/sec

(✱) PROBLEM 4.13 Retaining Wall

The below retaining wall is subjected to earth excitement with a Peak Ground Acceleration $PGA = 0.4g$ which generates a horizontal acceleration coefficient of $k_h = 0.2$ (*).

The soil properties behind the wall are:

- Soil type C
- Soil friction angle $\emptyset = 35°$
- Friction angle between soil and wall $\delta = 18°$
- Soil unit weight = 120 pcf

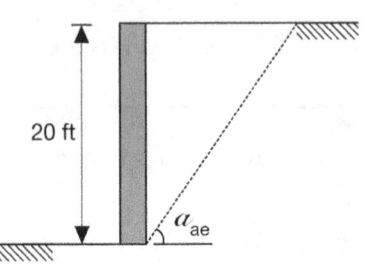

[Figure: Retaining wall 20 ft tall with failure wedge at angle α_{ae}]

Based on the above information, the pseudo static seismic active pressure coefficient k_{AE}, and the inclination of the seismic active wedge α_{ae} are most nearly:

(A) $k_{AE} = 0.2$ & $\alpha_{ae} = 33°$

(B) $k_{AE} = 0.4$ & $\alpha_{ae} = 32°$

(C) $k_{AE} = 0.4$ & $\alpha_{ae} = 49°$

(D) $k_{AE} = 0$ & $\alpha_{ae} = 53°$

(*) See solution section for information on how to calculate k_h when this coefficient is not provided in the question.

PROBLEM 4.14 Seismic Earth Pressure

The following statement(s) are true when it comes to lateral earth pressure seismic considerations:

I. Vertical acceleration in a pseudo-static calculation can be ignored as it does not occur at the same time the horizonal acceleration occurs, and it has a lesser effect.

II. Vertical acceleration should be combined with the horizontal ground acceleration as they occur simultaneously.

III. Generally, seismic analysis is not required when the slope of fill behind the retaining wall is flat, and the Peak Ground Acceleration is very low.

(A) I + II + III

(B) II + III

(C) I + III

(D) I + II

PROBLEM 4.15 Resultant of the Vertical Pseudo Static Forces

The weight of the failure wedge behind a retaining wall is $W_S = 7.5\ kip$ per linear foot and the vertical coefficient of vertical acceleration is $k_v = 0.1$. The vertical resultant for static and pseudo static forces behind this wall per linear foot is most nearly:

(A) $6.75\ kip\ \downarrow$

(B) $6.75\ kip\ \uparrow$

(C) $7.75\ kip\ \downarrow$

(D) $8.75\ kip\ \uparrow$

SOLUTION 4.1

Reference is made in this question to *FHWA NHI-11-032 LRFD Seismic Analysis and Design of Transportation Geotechnical Features and Structural Foundations Reference Manual,* Section 4.2.4 Depth to Bedrock.

For seismic analysis, the ideal soil profile should extend to a competent bedrock that has a shear wave velocity of $2,500\ ft/s$. **If reaching a competent bedrock is not feasible, the soil properties should be characterized over a depth of at least $200\ ft$.**

Correct Answer is (D)

SOLUTION 4.2

Reference is made in this question to *FHWA NHI-11-032 LRFD Seismic Analysis and Design of Transportation Geotechnical Features and Structural Foundations Reference Manual,* Section 4.4.4 Shear Wave Velocity.

The cross-hole method provides the best accuracy among all methods mentioned.

The below sketch illustrates the cross-hole, down-hole, and up-hole methods. The letter 'S' represents the wave source while 'R' represents the receiver. Knowing the distance between them allows us to calculate the wave speed.

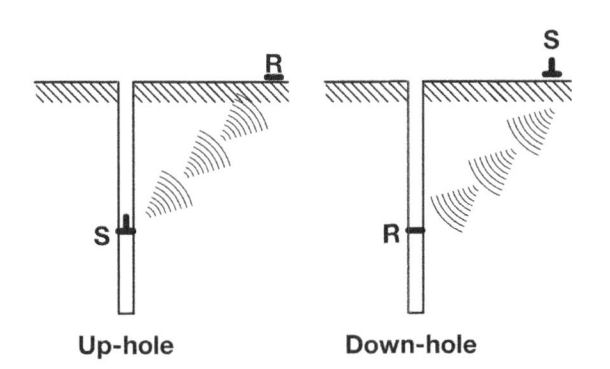

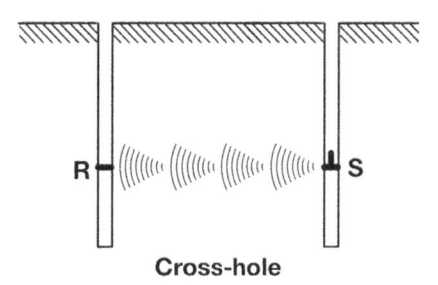

Seismic refraction, on the other hand, does not require intrusive measures or the erection of boreholes, and is better suited for measuring the depth of bedrock. It may not yield as accurate results as the previous methods when used to measure wave speeds.

Correct Answer is (A)

SOLUTION 4.3

Reference is made in this question to *FHWA NHI-11-032 LRFD Seismic Analysis and Design of Transportation Geotechnical Features and Structural Foundations Reference Manual,* Section 4.2.5 Cyclic Stress-Strain Parameters, and Table 4.5 (correlations).

Given no plasticity in this soil with $PI = 0$, this indicates the soil is cohesionless, hence, Imai & Tonouchi (1982) equation can be used from that abovementioned table as follows:

$$G_{max} = 15,560\ (N_{60})^{0.68}$$
$$= 15,560\ (15)^{0.68}$$
$$= 98,118.7\ kPa$$

Correct Answer is (A)

SOLUTION 4.4

Reference is made in this question to *FHWA NHI-11-032 LRFD Seismic Analysis and Design of Transportation Geotechnical Features and Structural Foundations Reference Manual,* Section 9.6.2 Effective Footing Dimensions, and Section 9.6.4 Bearing Capacity (*).

The equivalent uniform load is calculated using the effective or equivalent footing dimensions as presented in Figure 9-10 of the above reference, also depicted here for this example:

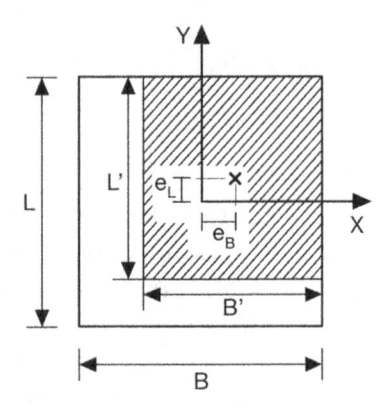

$$e_B = \frac{M_y}{V} = \frac{20 \; kip.ft}{80 \; kip} = 0.25 \; ft$$

$$e_L = \frac{M_x}{V} = \frac{40 \; kip.ft}{80 \; kip} = 0.5 \; ft$$

$$B' = B - 2e_B = 6 - 2 \times 0.25 = 5.5 \; ft$$

$$L' = L - 2e_L = 6 - 2 \times 0.50 = 5.0 \; ft$$

$$A' = B' \times L' = 5.5 \times 5.0 = 27.5 \; ft^2$$

$$\sigma' = \frac{80 \; kip}{27.5 \; ft^2} = 2.9 \; ksf$$

Correct Answer is (C)

(*) You can also refer to the *NCEES Handbook version 2.0,* Section 3.4.2.2 Eccentric and inclined Loaded Footings.

SOLUTION 4.5

Reference is made to the *ASCE 7-16* code, Section 19.2 SSI Adjusted Structural Demands, also to Table 12.2-1 of the same code to choose the Response Modification Factor R from.

$R = 3$ for the type of frame provided in this question – i.e., steel and concrete composite ordinary braced frame

$$V' \geq \alpha \, V$$

Where $\alpha = 0.7$ for $R \leq 3$ per Equation 19.2-3.

$$V' \geq 0.7 \times 400 = 280 \; kip$$

Correct Answer is (A)

SOLUTION 4.6

Reference is made to the ASCE 7-16 code, Section 20.4 Definitions of Site Class Parameters, and Table 20.3-1 Site Classification.

Per the above, the following equation is used and is only applied to the first 100 ft while the given borehole is 105 ft deep:

$$\bar{N} = \frac{\sum_{i=1}^{n} d_i}{\sum_{i=1}^{n} \frac{d_i}{N_i}}$$

Where is d_i the layer thickness and N_i is number of blows per foot. Also, you can refer to Solution 1.5 for more information on SPT test methods.

$$\bar{N} = \frac{30 + 45 + 25}{\frac{30}{8} + \frac{45}{30} + \frac{25}{55}} = 17.5$$

In reference to Table 20.3-1, and with $\bar{N} = 17.5$, this site is classified as **Site D – Stiff Soil**.

Correct Answer is (B)

SOLUTION 4.7

Using the ASCE 7-16 code Section 11.4.5 Design Spectral Acceleration Parameters.

S_{DS} is the design spectral response acceleration parameter in the short period range $= 2/3\, S_{MS}$. Per Eq. 11.4-1:

$$S_{MS} = F_a S_s$$

S_s is the mapped MCE_R Spectral response acceleration parameter at short periods, while MCE_R is the maximum considered earthquake $= 0.43$ (given).

F_a is the site coefficient and can be looked up from Table 11.4-1 of the ASCE 7-16 for site class C as follows:

$$S_s = 0.43,\ F_a = 1.3$$

$\rightarrow S_{DS} = (2/3) \times 1.3 \times 0.43 = 0.37$

The Seismic Design Category can now be determined from Table 11.6-1 of chapter 11 of ASCE 7-16. With $0.33 < S_{DS} < 0.5$ and a risk category II (retail) – see Table 1.5-1 from the reference, **the Seismic Design Category is mapped as C**.

Correct Answer is (C)

SOLUTION 4.8

Reference is made to *FHWA NHI-11-032 LRFD Seismic Analysis and Design of Transportation Geotechnical Features and Structural Foundations Reference Manual*, Section 6.3 Soil Liquefaction Hazard, page 6-21 defines **highly sensitive soils as those with Liquid limit < 0.4, Liquidity index > 0.6 and $(N_1)_{60} < 5$, and these applies to Soil A**.

Correct Answer is (A)

(⁂) SOLUTION 4.9

This solution explains the use of geophysics and seismic waves to estimate the portability of liquefaction for a certain sand layer with the use of *FHWA NHI-11-032 LRFD Seismic Analysis and Design of Transportation Geotechnical Features and Structural Foundations Reference Manual*, Section 6.3.3 Empirical and Numerical Evaluation Procedures for Liquefaction Potential, along with the following figures:

- Figure 6-17: Shear Wave Velocity - Liquefaction Resistance Correlation Chart.
- Figure 6-18: Soil Flexibility Factor (r_d) Versus Depth Curve.

The Cyclic Stress Resistance factor CSR is calculated using the following equation which can be found on page 6-36 of the above reference:

$$CSR = 0.65 \left(\frac{a_{max}}{g}\right)\left(\frac{\sigma_{vo}}{\sigma'_{vo}}\right) r_d$$

Where:

a_{max} is the Peak Ground Acceleration PGA given in the question as $0.3g$, where (g) is the gravitational acceleration.

σ_{vo} is the initial overburden stress and (σ'_{vo}) is the effective overburden stress, both calculated below. You can also refer to Solution 2.2 for further information on how to calculate the effective stress:

$$\sigma_{vo} = \sigma_{sand} = 120\, pcf \times 15\, ft$$
$$= 1{,}800\, psf\,(1.8\, ksf)$$

$$u_{sand} = 62.4\, pcf \times 10\, ft$$
$$= 624\, psf\,(0.62\, ksf)$$

$$\sigma'_{vo} = \sigma'_{sand} = \sigma_{sand} - u_{sand}$$
$$= 1.8 - 0.62 \cong 1.2\, ksf$$

r_d is the flexibility factor which can be determined using Figure 6-18 mentioned above.

As observed from this figure, (r_d) decreases with depth, which in turn lowers the CSR, thereby reducing the potential for liquefaction. Liquefaction is more likely to occur in shallow layers compared to deeper ones.

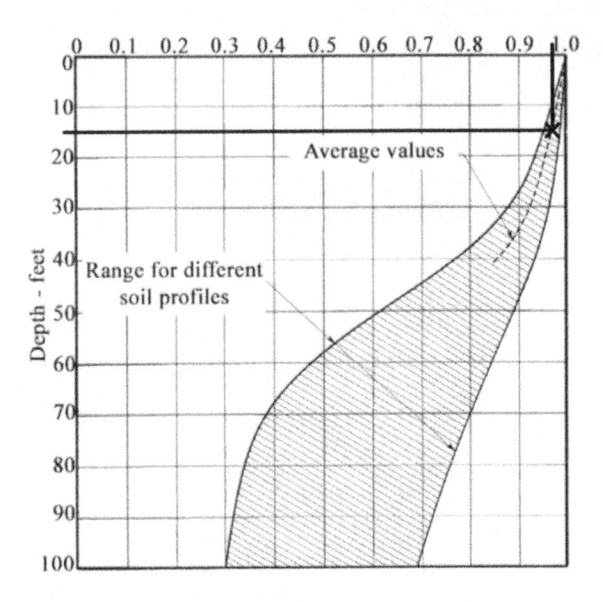

From the above $r_d = 0.98$.

Substituting all the above information into the CSR equation gives us the following result:

$$CSR = 0.65 \left(\frac{0.3g}{g}\right)\left(\frac{1.8}{1.2}\right) \times 0.98 = 0.29$$

Plotting a CSR value of 0.29 along with the corrected shear wave velocity of $v_{s1} = 150 \, m/sec$ on Figure 6-17 of the referenced manual as shown below (figures copied from FHWA originally provided by Andrus & Stokoe 2000 and Idriss 1971) indicates **a high potential for liquefaction in this soil layer at that depth during an earthquake with a magnitude of $M_w = 7.5$**.

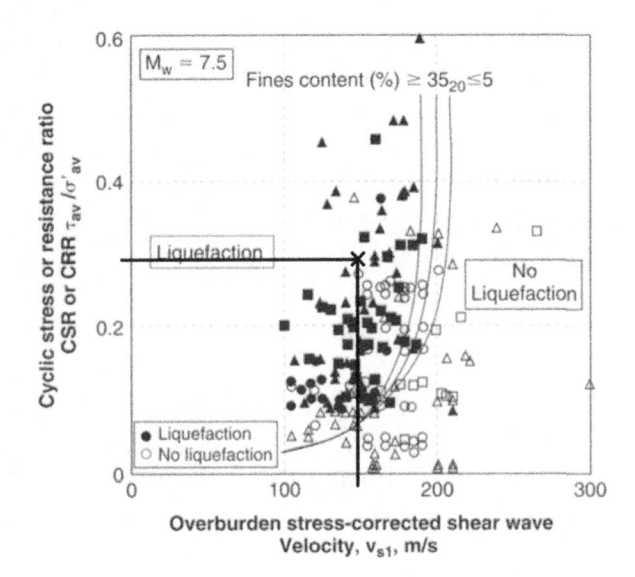

Correct Answer is (A)

(*) Some more information can be calculated from the above solution such as the induced cyclic shearing stress (τ_{ave}) which can be calculated as follows:

$$CSR = \frac{\tau_{ave}}{\sigma'_{vo}}$$

$$\tau_{ave} = CSR \times \sigma'_{vo}$$

$$= 0.29 \times 1.2$$

$$= 0.35 \, ksf$$

The above value indicates that any shear stress caused by seismic excitement, or any other lateral excitement, exceeding 0.35 ksf can cause liquefaction in this layer.

Based on this information, a safety factor for liquefaction can be calculated and further used to improve the design of foundations and structures.

(∗) **SOLUTION 4.10**
Reference is made in this question to *FHWA NHI-11-032 LRFD Seismic Analysis and*

Design of Transportation Geotechnical Features and Structural Foundations Reference Manual, Section 6.2.2 Limit Equilibrium Pseudo Static Stability Analysis, and Section 3.3.8 Local Site Effects.

For slope seismic stability analysis, it is essential to evaluate all potential circular and sliding wedges along with their failure mechanisms. To achieve this, the Capacity to Demand (C/D) ratio shall be determined. This involves determining seismic loads or moments and their corresponding resisting movements, in a similar fashion to an overturning assessment. The moment in this case shall be taken around the centroid of the failing circular surface as shown in the below schematic – the below schematic shows one of the failure surfaces only, a computer program normally assesses numerous similar failure surfaces to determine the one with the lowest, or most critical, C/D ratio (*):

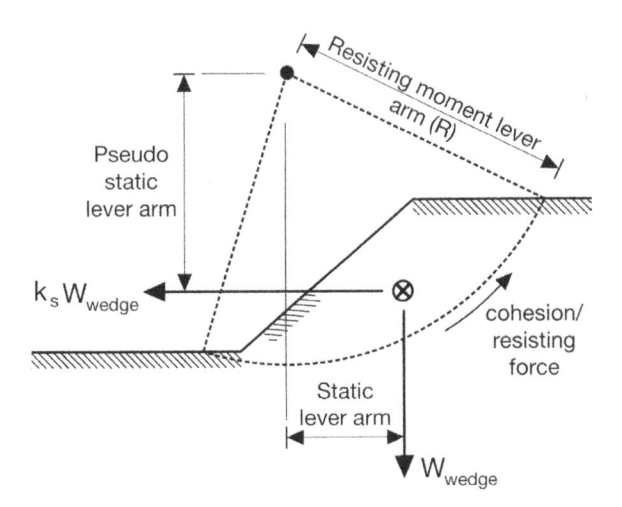

Pseudo static analysis is a widely used method for assessing the stability of slopes, retaining walls, and other geotechnical structures during seismic events. It simplifies the analysis by assuming that ground motion occurs instantaneously (statically) rather than over time, disregarding dynamic effects such as soil inertia.

In this approach, forces representing earthquakes are typically considered horizontal, neglecting vertical forces. To achieve this, the seismic coefficient k_s requested in this question is multiplied by the weight of each slice or wedge.

Based on this, the following equations from the referenced manual are used as follows:

$k_s = 0.5\alpha F_{pga} PGA$ ....Equation 6-5

$\alpha = 1 + 0.01H(0.5\beta - 1)$ ......Equation 6-3

$\beta = F_v S_1 / k_{max}$ ....Equation 6-4

$k_{max} = F_{pga} PGA$ ....Equation 6-1

Where:

F_{pga} is a site-specific factor and is collected from Table 3-6 of the reference used here as 1.2.

F_v is AASHTO's site factor and can be taken from Table 3-8 of the same reference as 2.0.

α is the height reduction factor.

Based on the above, those equations can be solved as follows:

$k_{max} = 1.2 \times 0.3 = 0.36$ ....Eq. 6-1

$\beta = \frac{2.0 \times 0.2}{0.36} = 1.11$ ....Eq. 6-4

$\alpha = 1 + 0.01 \times 65(0.5 \times 1.11 - 1) = 0.71$

$k_s = 0.5 \times 0.71 \times 1.2 \times 0.3 = 0.13$ Eq. 6-5

Correct Answer is (A)

(*) This schematic depicts both the driving forces (the "demand") with their lever arms – i.e., the static and the pseudo static – along with cohesion which plays part in the resisting forces (or moment around the identified circular centroid – the "capacity").

SOLUTION 4.11

Reference is made to the ASCE 7-16 code, Section 20.4 Definitions of Site Class Parameters, and Equation 20.4-1.

$$v_{s,av.} = \frac{\sum_{i=1}^{n} d_i}{\sum_{i=1}^{n} \frac{d_i}{v_{s,i}}}$$

$$= \frac{43 + 37 + 20}{\frac{43}{230} + \frac{37}{330} + \frac{20}{450}}$$

$$= 291.1 \, ft/sec$$

Correct Answer is (C)

SOLUTION 4.12

Reference is made in this question to *FHWA NHI-11-032 LRFD Seismic Analysis and Design of Transportation Geotechnical Features and Structural Foundations Reference Manual*, Section 4.3.3 Shear Wave Velocity.

From the above section, the small strain shear modulus G_{max} can be determined as follows:

$$G_{max} = \frac{E_{max}}{2(1+v)}$$

$$= \frac{2,500}{2(1+0.35)}$$

$$= 926 \, psi$$

The mass density of the soil (part of the shear wave velocity equation) is determined from Eq. 4-3 as follow:

$$\rho = \frac{\gamma_t}{g}$$

$$= \frac{120 \, \frac{lb}{ft^3}}{32.174 \, \frac{ft}{sec^2}}$$

$$= 3.73 \, \frac{lb.sec^2}{ft^4}$$

Based on this information, the shear wave velocity can be calculated – while taking close attention to units' conversion – using Equation 4-2 of the referenced manual as follows:

$$v_s = \sqrt{\frac{G_{max}}{\rho}}$$

$$= \sqrt{\frac{926 \, \frac{lb}{in^2} \times \left(\frac{144 \, in^2}{ft^2}\right)}{3.73 \, \frac{lb.sec^2}{ft^4}}}$$

$$= \sqrt{35,749.1 \, \frac{ft^2}{sec^2}}$$

$$= 189.1 \, ft/sec$$

Correct Answer is (B)

(✸) SOLUTION 4.13

Reference is made in this question to *FHWA NHI-11-032 LRFD Seismic Analysis and Design of Transportation Geotechnical Features and Structural Foundations Reference Manual*, Section 11.3.1 Mononobe-Okabe Seismic Earth Pressure Theory. The same can also be found in the *NCEES Handbook version 2.0* Section 3.1.5.

Start by checking Table 11-1 of the referenced manual which provides conditions when seismic analysis is not required, knowing that the slope angle above the wall is flat – i.e., $\beta = 0$.

F_{pga} in this table is a site-specific factor collected from Table 3-6 of the referenced manual $= 1.0$.

$$F_{pga}PGA = 1.0 \times 0.4 > 0.3$$

This result indicates that seismic analysis is required, and this precludes Option (D) from the solutions options.

The active pressure coefficient k_{AE} is computed using Equation 11.2 (*) from the referenced manual as follows:

$$k_{AE} = \frac{\cos^2(\emptyset-\theta-\omega)}{\cos\theta \cos^2\omega \cos(\theta+\omega+\delta^*)\left(1-\sqrt{\frac{\sin(\emptyset+\delta)\sin(\emptyset-\theta-\beta^*)}{\cos(\delta+\theta+\omega)\cos(\beta-\omega)}}\right)^2}$$

Where:
$\theta = tan^{-1}(k_h) = tan^{-1}(0.2) = 11.3°$ (**)

β is backfill slope $= 0$

$\emptyset$ is friction angle $= 35°$

δ is soil/wall friction angle $= 18°$

ω is wall batter (see Figure 11-11) $= 0$

$$k_{AE} = \frac{\cos^2(35-11.3-0)}{\cos 11.3 \cos^2 0 \cos(11.3+0+18)\left(1+\sqrt{\frac{\sin(35+18)\sin(35-11.3-0)}{\cos(18+11.3+0)\cos(0-0)}}\right)^2}$$

$$= \frac{\cos^2(23.7)}{\cos(11.3)\times 1 \times \cos(29.3)\left(1+\sqrt{\frac{\sin(53)\sin(23.7)}{\cos(29.3)\times 1}}\right)^2}$$

$$= \frac{0.84}{0.98 \times 1 \times 0.87 \left(1+\sqrt{\frac{0.8\times 0.4}{0.87 \times 1}}\right)^2}$$

$$\cong 0.40$$

This value can also be obtained from Figure 11-14 of the mentioned reference using a cohesion of $c = 0$, $k_h = 0.2$ and friction angle $35°$ (The same figure can be obtained from the *NCEES Handbook version 2.0* page 81).

The inclination of the seismic active wedge α_{ae} can be obtained from Figure 11-12 of the same reference FHWA NHI-11-032 page 11-13 using a backslope angle of *zero* as $\alpha_{ae} = 49°$. This figure is currently not provided by the NCEES Handbook.

Correct Answer is (C)

(*) The equation provided in the reference by FHWA NHI-11-032 is incorrect. It was cross-referenced with the original source (references C.[2] & C.[3] in the bibliography) and the necessary corrections made. The corrected angles in the equation are marked with an asterisk. On the other hand, the equation provided in the NCEES Handbook Section 3.1.5 is the correct equation.

In the FHWA NHI-11-032 reference, the incorrect equation yields a result of $k_{AE} = 0.34$. However, the correct equation gives a result of $k_{AE} = 0.4$. This discrepancy is further confirmed by examining Figure 11.14 which provides the correct answer.

It is important however to note that during the exam, you will not be penalized for errors in the provided references.

(**) k_h can be calculated using Section 11.4.1 Maximum Seismic Coefficient for Design with the use of ductility (r) and height adjustment factor (α) where the later was used in Solution 4.10 here.

SOLUTION 4.14
Reference is made in this question to *FHWA NHI-11-032 LRFD Seismic Analysis and Design of Transportation Geotechnical Features and Structural Foundations Reference Manual*, Section 11.3 Seismic Lateral Earth Pressures.

Statement	Comment
I. Vertical acceleration in a pseudo-static calculation can be ignored as it does not occur at the same time the horizonal acceleration occurs, and it has a lesser effect.	This statement is **True,** because of that we cannot superimpose the two forces.

II. Vertical acceleration should be combined with the horizontal ground acceleration as they occur simultaneously.	This statement is **Wrong** because the two accelerations are generally out of phase, and they should not be combined.
III. Generally, seismic analysis is not required when the slope of fill behind the retaining wall is flat, and the Peak Ground Acceleration is very low.	This statement is **True**, and it can be verified using Table 11.1 of the abovementioned reference.

Correct Answer is (C)

SOLUTION 4.15

Reference is made in this question to the *NCEES Handbook version 2.0* Section 3.1.5 Pseudo Static Analysis and Earthquake Loads, along with the figure which is provided in the same section showing the resultant (R) of static forces and seismic forces depicted as follows:

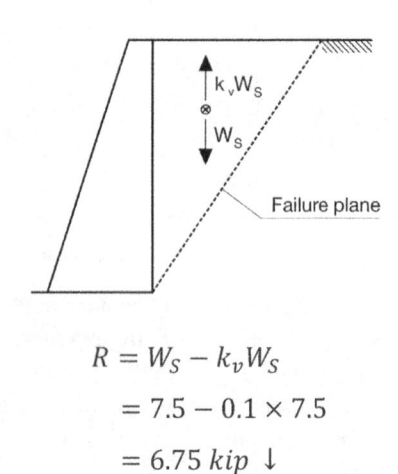

$$R = W_S - k_v W_S$$
$$= 7.5 - 0.1 \times 7.5$$
$$= 6.75 \; kip \downarrow$$

Correct Answer is (A)

V
EARTH STRUCTURES

Knowledge Areas Covered

SN	Knowledge Area
5	**Earth Structures, Ground Improvement, and Pavement** A. Ground improvement (e.g., grouting, soil mixing, preconsolidation/wick drains, lightweight materials, lime/cement stabilization, rigid inclusions, aggregate piers) B. Geosynthetic applications (e.g., separation, strength, filtration, drainage, reinforced soil slopes, internal stability of MSE) C. Slope stability evaluation and slope stabilization D. Embankments, earth dams, and levees (e.g., stress, settlement) E. Landfills and caps (e.g., interface stability, settlements, lining systems) F. Pavement and slab-on-grade design (e.g., rigid, flexible, porous, unpaved) G. Utility design and construction

PART V
Earth Structures

PROBLEM 5.1 *Subbase Stabilization*
A road contractor is requested to stabilize the subbase layer before the construction work starts as it includes moderate amount of clay gravel soils within its particles.

Considering that the *Plasticity Index* (PI) for the affected layer is '40', the best improvement method is as follows:

(A) Cement stabilization by mixing 3% Portland cement with the subbase material.

(B) Cement stabilization by mixing 9% Portland cement with the subbase material.

(C) Apply a small percentage of lime, typically 0.5% to 3%, to the affected material with a process called lime modification.

(D) Lime stabilization by mixing nearly 3% to 5% of lime to the affected layer.

(☆) PROBLEM 5.2 *Cement Stabilization*
The type of cement that shall be used for cement stabilization to provide better sulfate resistance is the following:

(A) Type I

(B) Type II

(C) Type III

(D) Type IV

PROBLEM 5.3 *Slope Stability/ Slope Safety Factor (1)*
The below slope belongs to an excavation in a soil with cohesion $c = 58\ psf$ and density $\gamma = 95\ pcf$ along with a friction angle of $\emptyset = 20°$.

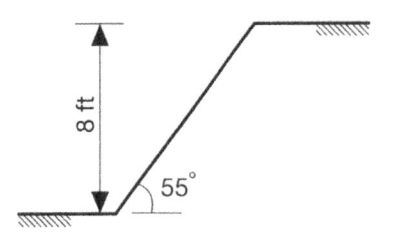

Using Taylor soil stability charts, the slope safety factor for this excavation is most nearly:

(A) 0.9

(B) 1.5

(C) 3.0

(D) 0.3

PROBLEM 5.4 *Consolidation Settlement*
The below graph plots the results of an odometer test for a clay sample where x-axis represents the logarithm of pressure, and y-axis is the void ratio (e).

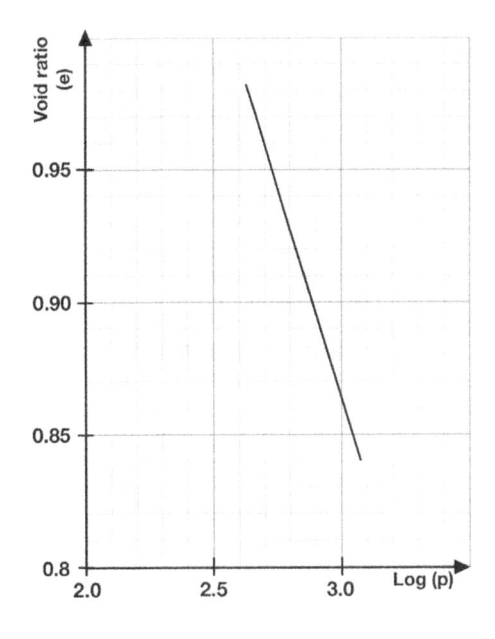

The expected settlement for a $4\ ft$ thick layer of this clay when pressure increases from an initial pressure of $500\ psf$ to a final pressure of $1,000\ psf$ is most nearly:

(A) 2.4 in

(B) 0.2 in

(C) 3.8 in

(D) 0.3 in

(✤) PROBLEM 5.5 *Base Layer Thickness*

A flexible pavement base layer has an expected serviceability loss $\Delta PSI = 1.0$ along with an $ESAL = 5 \times 10^6$.

Reliability $R = 90\%$ with a standard error $S_o = 0.3$.

Resilience modulus $M_r = 10,000\ psi$, and Elastic Modulus $E = 2 \times 10^5\ psi$ at $70°\ F$.

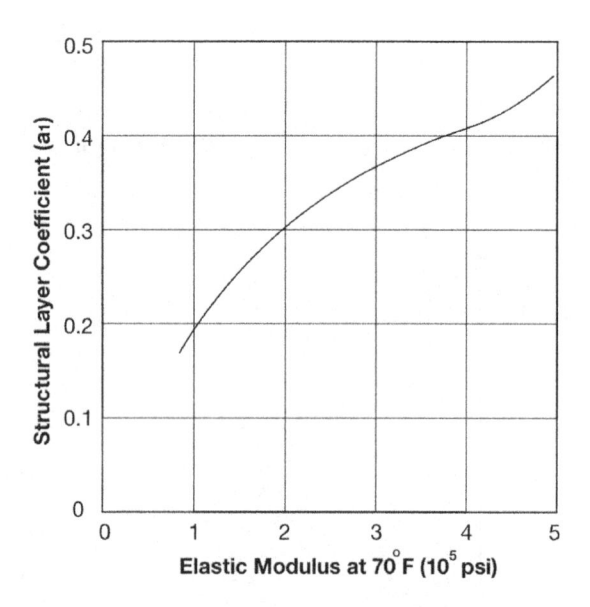

Based on the above information, the thickness of this layer is most nearly:

(A) 10 in

(B) 15 in

(C) 20 in

(D) 25 in

PROBLEM 5.6 *Distresses in Flexible Pavements*

The cause of a series of closely spaced ridges and valleys (ripples) in flexible pavements that are perpendicular to the traffic direction is usually caused by:

I. Insufficient base stiffness and strength.
II. Insufficient subgrade stiffness and strength.
III. Moisture and drainage problems.
IV. Freezing and thawing.

(A) I

(B) I + II

(C) I + II + III

(D) I + II + III + IV

PROBLEM 5.7 *Consolidation Settlement*

An $8\ ft$ thick clay layer lies between two sand layers – the top sand layer will be added at a later stage. The initial stress on the clay layer is $20\ psi$, which will increase to $35\ psi$ after adding the top sand layer.

To determine the time for the clay to settle by $1.8\ in$ due to this added stress, a $2\ in$ thick sample from this layer was tested in the lab. The void ratio of this sample changed from 0.9 at $20\ psi$ to 0.83 at $35\ psi$, and it took the sample 5 minutes to reach 50% consolidation while allowed to drain from its two ends to resemble site conditions.

Based on this information, the time required for the clay layer to settle by the $1.8\ in$ in the field is most nearly:

(A) 8 days

(B) 15 days

(C) 23 days

(D) 33 days

PROBLEM 5.8 Rapid Drawdown

The following statement(s) are true regarding the stability of the slope of a dam or embankment that holds water when it experiences a sudden drawdown of the water level:

I. The pore water pressure inside the embankment remains high which can cause slope instability.
II. The stabilizing forces (i.e., the external water) is removed which leads to slope instability.
III. As water seeps outside the embankment, effective stress increases which helps in stabilizing the slope.
IV. The base of the embankment could help if it was made of pervious material.

(A) I + III + IV
(B) II + III
(C) I + II + III
(D) I + II + IV

PROBLEM 5.9 Slope Stability (2)

The below sketch presents the details of a $45°$ degree sloped embankment with *zero* friction angle, cohesion $c = 450\ psf$, and a soil unit weight of $105\ pcf$.

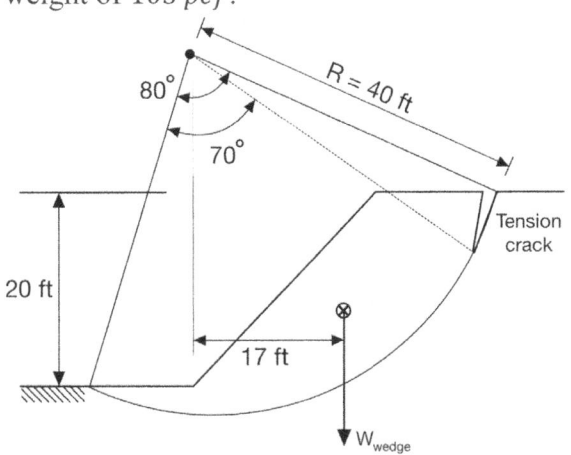

The presented arc represents a proposed trial failing plane with a radius of $40\ ft$, and weight of the proposed failing wedge of $W_{wedge} = 25\ kip$ per linear feet.

Based on the above information, the slope stability Factor of Safety for this trial wedge is most nearly:

(A) 2.1
(B) 2.6
(C) 0.6
(D) 0.4

PROBLEM 5.10 Slope Stability (3)

A very long slope, inclined at 25 *degrees* and $8\ ft$ deep, composed of soil with a density of $110\ pcf$ and an initial pore water pressure of $0.35\ ksf$, will experience a change in its slope stability Factor of Safety when all its water seeps away, and pore water pressure drops to *zero*, by most nearly:

(A) Improvement by 50%
(B) Reduction of 40%
(C) Improvement by 100%
(D) Reduction of 50%

PROBLEM 5.11 Slope Improvement

A new luxury apartment complex is being constructed on a hillside with a steep slope. The site is prone to landslides and soil erosion, especially during heavy rains. The project requires a solution that can provide strong lateral support, stabilize the slope, and ensure the safety of the buildings and residents. The chosen method should blend well with the natural environment and be cost-effective.

The best method in this case is (select two that applies):

- ☐ Soldier piles and lagging wall
- ☐ Anchored walls
- ☐ Geosynthetic Reinforced Soil (GRS) walls
- ☐ Soil nailing
- ☐ Mechanically Stabilized Earth (MSE) walls

PROBLEM 5.12 *MSE Internal Stability*
The below retaining wall is a 20 ft high geotextile reinforced backfilled with $\gamma = 120\ pcf$ and $\emptyset = 35°$ granular material.

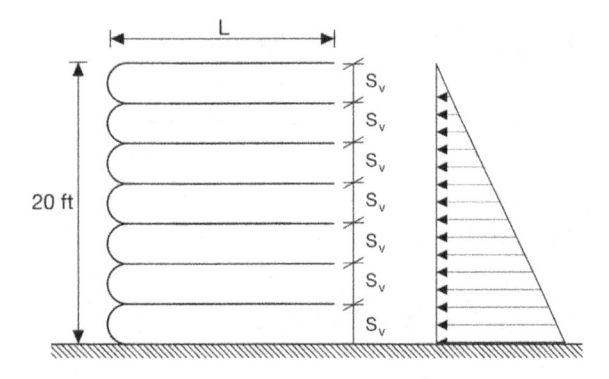

With an allowable tensile strength for the geotextile of $4.0\ kip/ft$, safety factor of 1.5, a shearing angle between the backfill and the geotextile of $\delta = 25°$ and adhesion between the geotextile and the backfill of $750\ psf$. The total geotextile length L and the lift thickness S_v should be:

(A) $L = 14\ ft, S_v = 2.85\ ft$
(B) $L = 10\ ft, S_v = 4\ ft$
(C) $L = 8\ ft, S_v = 6.7\ ft$
(D) $L = 6\ ft, S_v = 8\ ft$

PROBLEM 5.13 *Highway Improvement*
A new highway is being constructed over a region known for its soft, marshy terrain. The engineering team is concerned about the longevity and stability of the road due to the potential mixing of the high-quality aggregate base with the unstable subgrade soil under the stress of heavy traffic.

Given the risk of mixing between the aggregate base and the soft subgrade, the following geosynthetic materials could be used (select two that applies):

- ☐ Geogrids
- ☐ Geosynthetic Clay Liner (GCL)
- ☐ Geocell
- ☐ Geotextile
- ☐ Geomembrane
- ☐ Geocomposites

PROBLEM 5.14 *Landfill Design*
A site previously utilized for the disposal of oils, by-products, chemicals, and various wastes requires remediation.

The landfill's design incorporates several key features as shown in the below cross section: a reinforced layer to mitigate settlement due to the softening of the ground from all the disposals, a system to capture gaseous emissions using perforated pipes, and a barrier to prevent rainwater infiltration, which could intensify reactions within the existing soil.

Additionally, a gravel layer and, potentially, a soil cover will serve as the final top layer.

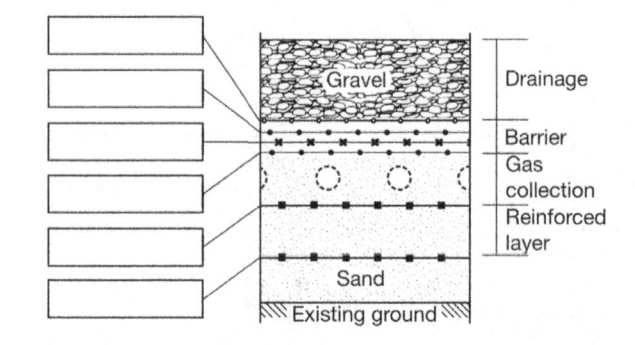

Given the available membranes listed below, indicate their respective placements within the above landfill cross-section. You may allocate any membrane to multiple locations as needed:

- Geogrid
- Geosynthetic Clay Liner (GCL)
- Geotextile
- Geomembrane

PROBLEM 5.15 *Consolidation Wicks*
An embankment is constructed for a road project that has an alluvial nature and is expected to settle causing geometry and stability concerns.

To address this, the installation of Prefabricated Vertical Drains (PVDs), known as wicks, in conjunction with preloading with a surcharge load has been identified as a cost-effective strategy.

Given this context, identify the true statement(s) from the below:

I. The above solution can lead to increasing the required preloading surcharge load.
II. The rate of strength gain is improved using this method.
III. This method eliminates post construction settlement in the embankment.
IV. For better results, water should be reduced or eliminated as much as possible prior to applying those wicks and the surcharge load.

(A) I + III + IV

(B) II + III

(C) II

(D) I + II + IV

PROBLEM 5.16 *Lightweight Subgrade*
In order to reduce the lateral load on a retaining wall which supports an approach embankment for a transportation project, the subgrade layer has been replaced with geofoam per the below section.

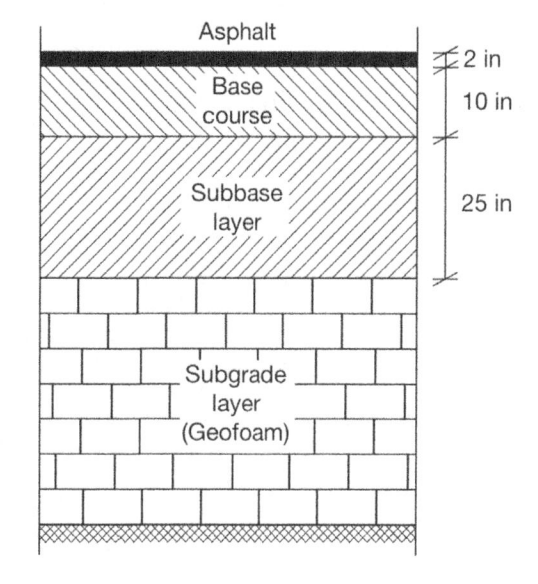

Knowing that the density of the wearing course (i.e., the asphalt) is 135 pcf, the base course's density is 125 pcf and the subbase is 120 pcf. Also, ignoring the weight of the geofoam layer, and accounting for traffic on the embankment of 200 psf, the best type of polystyrene geofoam to be used at 1% strain in this case is:

(A) EPS15

(B) XPS20

(C) EPS12

(D) EPS19

PROBLEM 5.17 *Vibrocompaction*
The below gradation represents a soil that is being considered for deep vibrocompaction in a bid to reduce its liquefaction potential and improve its shear strength.

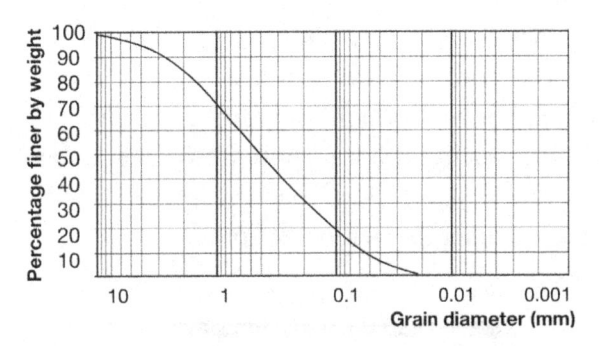

Based on this gradation, the following applies to this soil during this operation:

(A) Vibrocompaction has no effect on this soil and better to use stone columns instead.

(B) Vibrocompaction works but requires sand backfill during the operation.

(C) Vibrocompaction works but requires gravel fill during the operation.

(D) Vibrocompaction works and does not require any fill during the operation.

PROBLEM 5.18 *Supported Embankments*

The following statement(s) are true for Column Supported Embankments CSEs:

I. CSEs are vertical columns that reinforce embankments by uniting weak soil strata, without necessarily extending to a stiff soil layer.
II. Non-compressible columns are used for weak soils that cannot support surface loads.
III. Embankments punching shear is of a concern when designing for CSEs.
IV. Using CSEs may lead to project delays, making consolidation wicks a preferred option.

(A) I + III + IV

(B) II + III

(C) II

(D) I + II + IV

PROBLEM 5.19 *Bitumen Stabilization*

A construction project requires the stabilization of a subgrade layer to be used as a detour. The subgrade embankment consists of a high percentage of fine aggregate and material passing sieve No. 200. The project team is evaluating four different types of asphalt emulsions to be used on this layer to stabilize this layer and provide dust control.

The best option to be used for this purpose is:

(A) Rapid Curing Asphalt (RC-250)

(B) Medium Curing Asphalt (MC-3000)

(C) Medium Setting Asphalt (CMS-2)

(D) Slow Setting Asphalt (SS-1)

PROBLEM 5.20 *Edgedrain Construction*

The best method to prevent the perforated pipe for an edgedrain of a road from being crushed by machinery and heavy truck loads during construction is the following:

(A) Replace the perforated pipe with a rigid concrete pipe.

(B) Install the edgedrain after pavement construction.

(C) Replace the drain with a daylighted permeable base layer.

(D) Place the pipe deeper with more fill on top of it.

SOLUTION 5.1

Cement stabilization is considered when *plasticity index* is less than 10 (**) and is used to strengthen granular soils by mixing in Portland cement, typically 3 – 5% of the soil dry weight. Check *NCEES Handbook* Section 2.5.4 for more information.

Lime (*) modification is used to improve fine grained soils with the addition of 0.5% *to* 3% to the soil dry weight.

However, with a plasticity index > 10 for subbase or base materials with moderate and "predominant" amount of clay gravel soils, **lime stabilization is considered the best option**. Typically, 3% *to* 5% should be enough to dry up the mud contained in the subbase layer.

Correct Answer is (D)

(*) You can find more information on lime stabilization and admixtures stabilization in *FHWA NHI-05-037 Geotechnical Aspects of Pavements*, Section 7.6.5 Admixture Stabilization and page 7-79.

(**) In FHWA NHI-05-037 the same information is mentioned on page 7-84 with the plasticity index should be less than 20 to qualify for cement stabilization.

SOLUTION 5.2

Reference is made to *FHWA NHI-05-037 Geotechnical Aspects of Pavements*, Section 7.6.5 Admixture Stabilization, and page 7-84.

Type II cement is the type that shall be used to provide greater sulfate resistance.

Correct Answer is (B)

(⁂) SOLUTION 5.3

The *NCEES Handbook,* Chapter 3 Geotechnical, Section 3.6 Slope Stability is referred to.

The handbook provides two charts for Taylor (1948). The second is only applicable when friction angle $\phi = 0$ and a rock layer has been identified below the slope where the depth factor $D > 1$, which is not the case here.

The first chart – copied in the following page for ease of reference (used with permission from FHWA) – is used in this case.

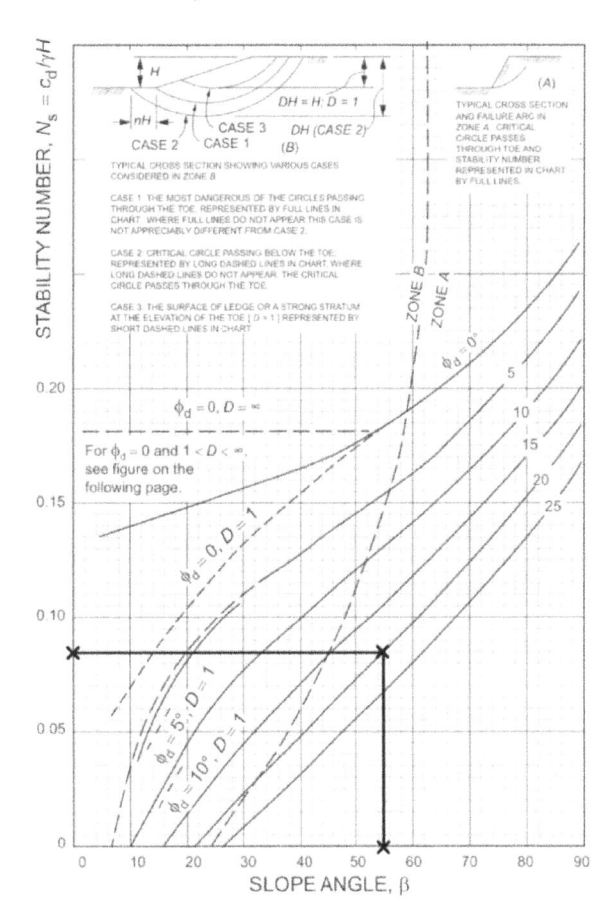

There are two factors that we shall define prior to performing the calculation:

> c_d is the developed, or mobilized, cohesion, which is the cohesion that develops at the slip surface upon failure.

ϕ_d is the developed, or mobilized friction angle, which is the friction angle that develops at the slip surface upon failure.

The safety factor requested in this question represents safety against forming a slip surface, in which case:

$$FS = F_c = \frac{c}{c_d}$$

Similarly, the safety factor for the friction angle shall be calculated and shall equal to:

$$FS = F_\phi = \frac{\tan \phi}{\tan \phi_d}$$

The above process is iterative in nature, and we may have to perform two or more iterations until the following equation is satisfied:

$$FS = F_\phi = F_c$$

Iteration 1: assume $\phi = \phi_d = 20°$

In reference to Taylor's (1948) first chart, using a slope angle $\beta = 55°$ – first iteration shown on the chart – the stability number $N_s = 0.085$.

$$N_s = \frac{c_d}{\gamma H}$$

$\rightarrow c_d = \gamma H N_s$

$= 95 \frac{lb}{ft^3} \times 8\ ft \times 0.085$

$= 64.6\ psf$

$F_c = \frac{c}{c_d} = \frac{58\ psf}{64.6\ psf} = 0.9 = F_\phi$

$\phi_d = \arctan\left(\frac{\tan \phi}{F_\phi}\right)$

$= \arctan\left(\frac{\tan 20°}{0.9}\right)$

$= 22°$

Iteration 2: assume $\phi_d = 22°$

In reference to Taylor (1948) chart, using interpolation $\rightarrow N_s = 0.077$.

$c_d = \gamma H N_s$

$= 95 \frac{lb}{ft^3} \times 8\ ft \times 0.077$

$= 58.5\ psf$

$F_c = \frac{c}{c_d} = \frac{58\ psf}{58.5\ psf} = 0.99 = F_\phi$

$\phi_d = \arctan\left(\frac{\tan \phi}{F_\phi}\right)$

$= \arctan\left(\frac{\tan 20°}{0.99}\right)$

$= 20°$

It is obvious at this stage that the safety factor falls somewhere between '0.9' to '1.0'.

Correct Answer is (A)

SOLUTION 5.4
The *NCEES Handbook,* Chapter 3 Geotechnical, Section 3.2.1 Normally Consolidated Soils is referred to.

Settlement in a clay layer is calculated using the following equation:

$$S_C = \sum_{1}^{n} \frac{C_c}{1 + e_o} H_o \log\left(\frac{p_f}{p_o}\right)$$

Where C_c is the compression index and is calculated using the slope of '$\log(p) - e$' graph shown below (*). e_o is initial void ratio and can be picked up from the graph by substituting for $\log(p_o)$. H_o is the layer thickness, p_f is the final pressure and p_o is the initial/original pressure. n in the equation represents the number of layers, in which case the question did not specify more than one layer.

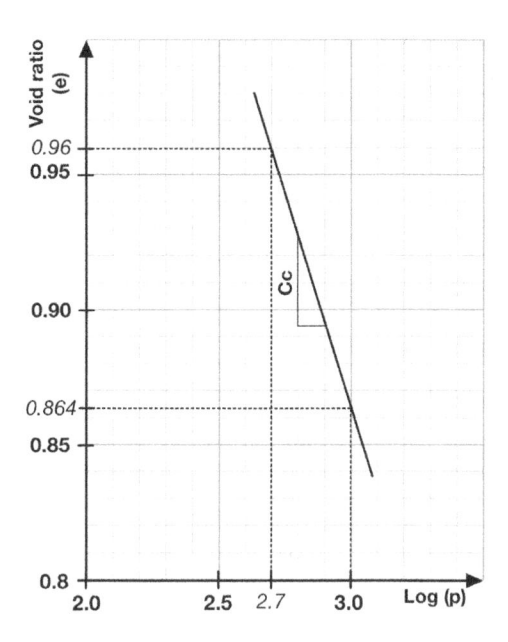

$$C_c = \frac{\Delta e}{\Delta \log(p)} \quad (*)$$

$$= \frac{0.96 - 0.864}{3.0 - 2.7}$$

$$= 0.32$$

$$e_o = e_{@ \, [\log(500) \, = \, 2.7]} = 0.96$$

$$S_C = \frac{C_c}{1+e_o} H_o \, Log \left(\frac{p_f}{p_o}\right)$$

$$= \frac{0.32}{1+0.96} \times 4 \, ft \times Log \left(\frac{1{,}000}{500}\right)$$

$$= 0.2 \, ft \, (2.4 \, in)$$

Correct Answer is (A)

(*) For the removal of doubt, C_c is an absolute value and is calculated using the delta of void ratio Δe at the numerator. The graph shown in the *NCEES Handbook* version *1.1* might be misleading and one may think that Δe should be at the denominator instead which is wrong, this is corrected in version *1.2* and *2.0* now.

(✻) **SOLUTION 5.5**
Reference is made to *FHWA NHI-05-037 Geotechnical Aspects of Pavements,* Appendix C.2 Flexible Pavement Structural Design.

This section, along with Chapter 5 & 6 of the same reference, define the Structural Number SN, thickness (D) and the layer coefficient (a) which is taken from the figure in the question for the base layer as $a_1 = 0.3$.

As we are attempting to compute the first layer's thickness, Structural Number SN_1 is used as follow:

$$SN_1 = a_1 D_1$$
$$= 0.3 D_1$$

The Structural Number SN_1 can be calculated by solving the following equation taken from the above reference:

$$log_{10}(W_{18}) = Z_R S_0 + 9.36 log_{10}(SN_1 + 1)$$

$$- 0.20 + \frac{log_{10}\left(\frac{\Delta PSI}{4.2 - 1.5}\right)}{0.40 + \frac{1094}{(SN_1 + 1)^{5.19}}}$$

$$+ 2.32 log_{10}(M_R) - 8.07$$

Where the following attributes were given in the question:

$W_{18} = 5 \times 10^6$

$\Delta PSI = 1.0$

$S_o = 0.3$

$Z_R = -1.282$ using Table C.3 & $R = 90\%$

$M_r = 10{,}000 \, psi$

Substituting this information into the above equation gives the following equation:

$$6.07 = 9.36 log_{10}(SN_1 + 1) + \frac{-0.43}{0.40 + \frac{1094}{(SN_1 + 1)^{5.19}}}$$

This is a complex equation that cannot be solved during the time given during the exam, however, we can solve it by iteration using the options given in the question as follows:

Option (A):

$SN_1 = a_1 D_1 = 0.3 \times 10 = 3$

Substitute it in the equation:

$9.36 \log_{10}(3+1) + \dfrac{-0.43}{0.40 + \dfrac{1094}{(3+1)^{5.19}}} = 5.28$

Option (B):

$SN_1 = a_1 D_1 = 0.3 \times 15 = 4.5$

Substitute it in the equation:

$9.36 \log_{10}(4.5+1) + \dfrac{-0.43}{0.40 + \dfrac{1094}{(4.5+1)^{5.19}}} = 6.16$

Option (C):

$SN_1 = a_1 D_1 = 0.3 \times 20 = 6$

Substitute it in the equation:

$9.36 \log_{10}(6+1) + \dfrac{-0.43}{0.40 + \dfrac{1094}{(6+1)^{5.19}}} = 6.94$

Option (D):

$SN_1 = a_1 D_1 = 0.3 \times 25 = 7.5$

Substitute it in the equation:

$9.36 \log_{10}(7.5+1) + \dfrac{-0.43}{0.40 + \dfrac{1094}{(7.5+1)^{5.19}}} = 7.67$

The closest of the above four options is option B with a thickness of $D_1 = 15\ in.$ (*)

Correct Answer is (B)

(*) To confirm the correctness of the above method, this question was also solved using the original AASHTO reference which is not required in this exam (AASHTO's Guide for Design of Pavement Structures GDPS for the year 1993). This extra alternative solution is provided here for information only.

The following graph is copied from Chapter 3, Table 3.1 of AASHTO 1993 – Design chart for Flexible Pavements – used here with permission from AASHTO:

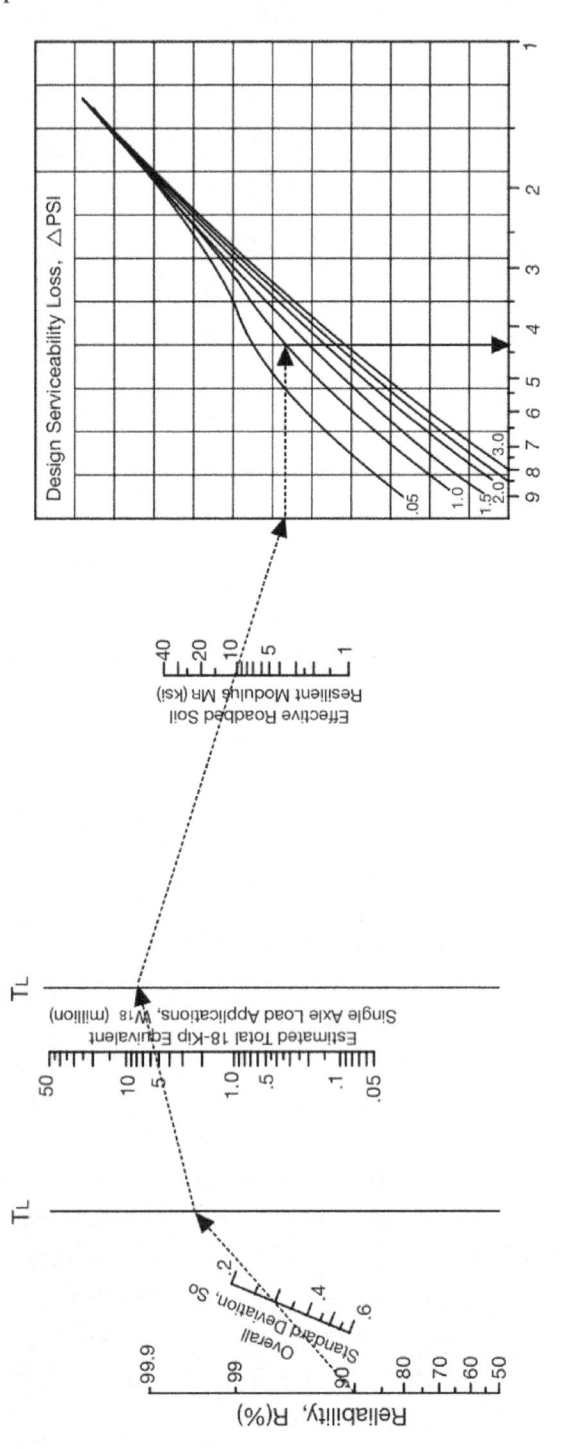

The above chart is used to determine the Structural Number SN_1. To obtain this, we start from the left most of the nomograph using reliability $R = 90\%$ and extending a dotted arrow from this axis to the first turning line T_L passing through the standard deviation $S_o = 0.3$.

Connect the first intersection with the turning line T_L to the second turning line passing the dotted arrow through the ESAL axis at 5×10^6. From this point, connect the last arrow to the $\Delta PSI/SN$ chart passing through the M_r axis at $10\ ksi$.

Using the above chart, a Structural Number of $SN_1 = 4.4$ is obtained for this layer.

Applying the layer coefficient equation described previously:

$$SN_1 = a_1 D_1$$
$$4.4 = 0.3 \times D_1$$
$$\rightarrow D_1 = 14.7\ in$$

SOLUTION 5.6
The *NCEES Handbook,* Chapter 3 Geotechnical, Section 3.19 Pavements, is referred to.

The description given in the question is termed as Corrugation. Corrugation is caused by **insufficient base stiffness and strength**.

Correct Answer is (A)

SOLUTION 5.7
The *NCEES Handbook version 2.0,* Section 3.2.1 Normally Consolidated Soils, and Section 3.2.3 Time Rate of Settlement, are both used in this solution.

In order to determine the time required to achieve a certain settlement in the field, start with determining the maximum settlement for the required layer using the equations provided in the relevant sections of the NCEES Handbook. You can also refer to Solution 5.4 here.

Information to be determined from lab:

The compression index C_c should be determined first using lab results as follows:

$$C_c = \frac{\Delta e}{\Delta \log(p)}$$
$$= \frac{0.90 - 0.83}{\log(35) - \log(20)}$$
$$= 0.29$$

The coefficient of consolidation C_v should also be determined from lab results for later use. C_v relies on $T_{v@50\%} = 0.197$ (see table at page 85 of the NCEES Handbook version 2.0), and $H_d = 1\ in$ (not $2\ in$ given the sample is allowed to drain from its two sides):

$$C_v = \frac{T_{v@50\%} \times H_d^2}{t}$$
$$= \frac{0.197 \times (1\ in)^2}{5\ min}$$
$$= 0.0394\ in^2/min$$

Lab results applied to the layer on the field:

Based on the compression index C_c, the maximum settlement for the $8\ ft$ thick layer can be computed as follows:

$$S_c = \frac{C_c}{1+e_o} H_o\ Log\left(\frac{p_f}{p_o}\right)$$
$$= \frac{0.29}{1+0.9} \times 8\ ft \times Log\left(\frac{35}{20}\right)$$
$$= 0.3\ ft\ (3.6\ in)$$

The above information is used to determine the average degree of consolidation U%, based upon which, the coefficient $T_{v@U\%}$ can be interpolated from the table provided in page 85 of the NCEES Handbook (or the FHWA-NHI-06-088 manual page 7-32).

$$U\% = \frac{1.8\ in}{3.6\ in} = 50\%$$

$$\rightarrow T_v = 0.197$$

Apply the above to the equation used to determine C_v considering the thickness of the field layer should be halved. This adjustment is necessary because the layer is drained from both its top and bottom, being situated between two permeable layers:

$$t = \frac{T_{v@50\%} \times H_d^2}{C_v}$$

$$= \frac{0.197 \times \left(\frac{8\ ft \times \frac{12\ in}{1\ ft}}{2}\right)^2}{0.0394\ in^2/min}$$

$$= 11{,}520\ min\ (8\ days)$$

Correct Answer is (A)

SOLUTION 5.8

When a dam, or an embankment, that holds water experiences a sudden drawdown, the stability of its slopes can be significantly affected. See below why and how:

Pore Water Pressure:
During a sudden drawdown, the water level outside the embankment drops quickly, but the pore water pressure within the embankment remains high. This imbalance can reduce the effective stress in the soil, leading to a decrease in its shear strength and potentially causing slope instability.

This marks statement I as correct.

Stabilizing Forces:
The rapid reduction in external water pressure removes the stabilizing force on the upstream slope, making it more susceptible to failure. The soil may not have enough time to adjust to the new conditions, leading to a higher risk of slope failure.

This marks statement II as correct.

Seepage Forces:
As water drains out of the embankment, seepage forces can develop, which may further destabilize the slope. These forces are critical if the drawdown is faster than the rate at which water can escape from the soil pores.

This marks statement III as incorrect.

Base Material:
If the base material is pervious (e.g., sand or gravel), water can drain more quickly and vertically, reducing pore water pressure more rapidly. This can help stabilize the slope faster.

On the other hand. if the base material is impervious (e.g., clay), water drains more slowly, maintaining higher pore water pressures for a longer period. This can lead to prolonged instability and a higher risk of slope failure during rapid drawdown. Moreover, in this case water will seep to the sides creating lateral pressure causing further slope deterioration.

This marks statement IV as correct.

The correct statements in this case are I, II and IV.

Correct Answer is (D)

SOLUTION 5.9

You can refer to Solution 4.10 here as it discusses potential circular and sliding wedges along with their failure planes, and their static and pseudo static lever arms. In this question we are considering the static lever arm only which is $17\ ft$ in this case.

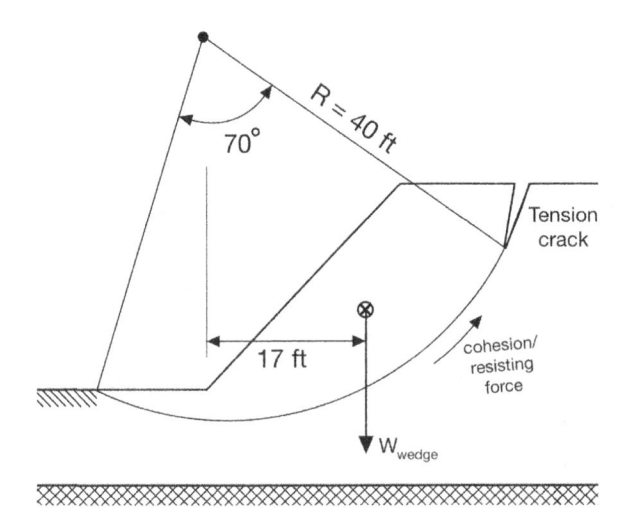

The sliding moment in this case is evaluated using the mass of the wedge alone, while the resisting moment is evaluated using cohesion multiplied by the failing surface ignoring the tension crack. See below:

$$M_{sliding} = 25\ kip/ft \times 17\ ft$$
$$= 425\ kip.ft/ft$$

$$M_{resisting} = c \times arc\ length \times lever\ arm$$
$$= c \times (\theta_{rad} \times r) \times r$$
$$= c \times \theta_{rad} \times r^2$$
$$= 0.45\ \frac{kip}{ft^2} \times \left(\frac{70}{180}\pi\right) \times (40\ ft)^2$$
$$= 879.2\ kip.ft/ft$$

$$FS = \frac{M_{resisting}}{M_{sliding}}$$
$$= \frac{879.2}{425}$$
$$= 2.1$$

Correct Answer is (A)

SOLUTION 5.10

The *NCEES Handbook version 2.0*, Section 3.6.4 Infinite Slope, is referred to. This information can also be found in *EM 1110-2-1902 USACE Engineering and Design: Slope Stability*, Section C-7 The Infinite Slope Method.

Factor of Safety with and without pore water pressure are calculated and divided by each other to determine improvement or reduction with the relevant percentage. See below:

$$FS_{pore\ pressure} = \frac{\tan\phi'}{\tan\beta}[1 - r_u(1 + \tan^2\beta)]$$

$$= \frac{\tan\phi'}{\tan\beta}\left[1 - \frac{0.35\ ksf}{0.11\ kcf \times 8\ ft}(1 + \tan^2 25)\right]$$

$$= \frac{\tan\phi'}{\tan\beta} \times 0.52$$

$$FS_{no\ pore\ pressure} = \frac{\tan\phi'}{\tan\beta}$$

$$\%age\ change\ in\ FS = \frac{FS_{no\ pore\ pressure}}{FS_{pore\ pressure}}$$

$$= \frac{\frac{\tan\phi'}{\tan\beta}}{\frac{\tan\phi'}{\tan\beta} \times 0.52}$$

$$= 1.92$$

The above indicates that there will be 100% improvement in the safety factor for the stability of the slope (*).

Correct Answer is (C)

(*) When comparing embankment slopes with and without pore water pressure, the one without pore water pressure is generally more stable. See below why:

Shear Strength: Pore water pressure reduces the effective stress in the soil, which in turn decreases its shear strength. Lower shear strength makes the slope more susceptible to failure.

Stability: High pore water pressure can lead to instability, especially in saturated soils. It can cause a reduction in frictional resistance, making the slope more prone to sliding.

Drainage: Proper drainage systems can help manage pore water pressure, but if not adequately designed, excess pore water can accumulate, leading to potential slope failure.

Also, check the brief provided in Solution 5.8 for more information on slope stability when pore water pressure is involved.

SOLUTION 5.11

☑ **Mechanically Stabilized Earth (MSE) walls**

Mechanically Stabilized Earth (MSE) walls provide strong lateral support and are effective in stabilizing slopes.

MSE walls are constructed using layers of compacted backfill and reinforcing elements, such as geogrids or steel strips, which provide excellent stability and support. The facing can be made of precast concrete panels, wire mesh, or other materials that blend well with the natural environment.

Moreover, MSE walls are cost-effective and can be used for retaining walls, bridge abutments, and embankments.

The method used to apply this MSE wall in this case involves the following: The process begins with clearing the area of vegetation and debris, followed by excavating the slope to create terraces or benches. A base layer of compacted granular material is then placed to provide a stable foundation. Geosynthetic materials, such as geogrids or geotextiles, are laid horizontally and alternated with compacted soil layers until the desired height is reached. The facing of the MSE wall can be covered with natural vegetation for erosion control and aesthetic integration, or with precast concrete panels for a more structured appearance as discussed previously.

Proper drainage should be provided to prevent water buildup behind the slope.

☑ **Geosynthetic Reinforced Soil (GRS) walls**

GRS walls use layers of geosynthetic materials (such as geotextiles or geogrids) combined with compacted soil to create a reinforced soil structure.

For the removal of doubt, GRS walls can be applied without removing and rebuilding the existing slope. The slope area is first cleared of vegetation and debris, and the slope is excavated to the required depth and shape. A base layer of compacted granular material is placed, followed by horizontal layers of geosynthetic materials (geotextiles or geogrids) and compacted soil, alternated until the desired height is reached. The facing of the GRS wall can be covered with natural vegetation for erosion control and aesthetic integration, or with precast concrete panels for a structured appearance.

Proper drainage should be provided to prevent water buildup behind the slope or wall.

- **Soldier pile and lagging wall**
  This method is typically used for temporary and permanent retaining walls in urban construction. While it provides good lateral support, it may not be the best choice for stabilizing a large, steep hillside prone to landslides and erosion. This method is more suited for vertical excavations rather than slope stabilization.

- **Anchored Walls**
  Anchored walls use tiebacks or anchors to provide additional support to retaining structures. While they are effective for deep excavations and high retaining walls, they may not provide the comprehensive slope stabilization needed for a hillside.

  Moreover, the installation of anchors can be complex and costly, and may not blend well with the natural environment.

- **Soil Nailing**
  Soil nailing involves inserting steel bars into a slope or excavation face and grouting them in place. While this method is effective for slope stabilization and retaining walls, it may not provide the same level of long-term stability and aesthetic integration as MSE or GRS walls. Soil nailing is also more suitable for smaller slopes and may not be as cost-effective for large hillside projects.

SOLUTION 5.12
The *NCEES Handbook version 2.0*, Section 3.13.5.1 Geotextile-Reinforced Walls (*), is referred to in this question.

Given there is no other pressures apart from the earth pressure, start with computing the Rankine active coefficient as follows:

$$K_a = \tan^2\left(45 - \frac{\phi}{2}\right)$$
$$= \tan^2\left(45 - \frac{35}{2}\right)$$
$$= 0.27$$

With the lateral pressure $\sigma_{hs} = K_a \gamma Z$, the equation for uplift thickness can be written as follows taking into account the direction of axis Z as shown in the following sketch:

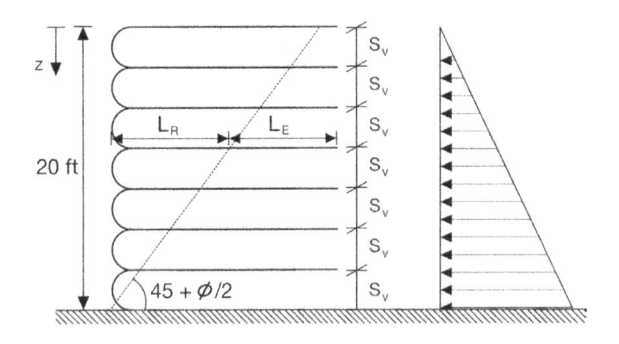

$$S_V = \frac{T_{allow}}{(K_a \gamma Z) FS}$$
$$= \frac{4.0 \, kip/ft}{(0.27 \times 0.12 \, kip/ft^3 \times Z) \times 1.5}$$
$$= \frac{82.3}{Z}$$

With numerous possible values of Z, the most critical one is the lowest giving a lift thickness of $4.1 \, ft$ occurring at $z = 20 \, ft$. Will use $4 \, ft$ considering this is one of the values given in the question options.

The total length of embedment L is calculated as follows:

$$L = L_R + L_E$$

With:

$$L_R = (H - Z)\tan\left(45 - \frac{\phi}{2}\right)$$

$$L_E = \frac{S_V (K_a \gamma Z) FS}{2(c_a + \gamma Z \tan(\delta))}$$

$$L = (H - Z)\tan\left(45 - \frac{\phi}{2}\right) + \frac{S_V (K_a \gamma Z) FS}{2(c_a + \gamma Z \tan(\delta))}$$

$$= (20 - Z)\tan(27.5) + \frac{4 \times (0.27 \times 0.12 \times Z) \times 1.5}{2(0.75 + 0.12 Z \tan(25))}$$

$$= 0.52(20 - Z) + \frac{0.0972 Z}{0.75 + 0.056 Z}$$

Try few values of Z at $4\,ft$ intervals and select the largest value of L as follows:

$$L_{Z=4} \approx 9\,ft$$
$$L_{Z=8} \approx 7\,ft$$
$$L_{Z=12} \approx 5\,ft$$
$$L_{Z=16} \approx 3\,ft$$
$$L_{Z=20} \approx 1\,ft$$

Use $L = 10\,ft$ and $S_V = 4\,ft$ as suggested in the question's option B.

Correct Answer is (B)

(*) You can find more information and extra equations on MSE walls in *UFC 3-220-10 Soil Mechanics,* Section 7-9 Mechanically Stabilized Earth Slopes.

For instance, you can find more equations relevant to pull out strength per geotextile layer ($P_r = F^* \alpha \sigma'_v L_e C$) and the interaction coefficient $\left(C_i = \frac{\tan(\delta)}{\tan(\phi')} \right)$ in Section 7-9.4 Soil-Geosynthetic Interaction.

📄 SOLUTION 5.13

This case study requires the separation between the two mentioned regions or layers. Separation includes placing a strong, porous barrier between two different layers to keep them working well without mixing up. For roads built on soft ground, this barrier stops the strong top layer from sinking into the soft bottom layer when heavy vehicles pass over. This method helps the road stay stable and last longer. Geotextiles are commonly used for this, especially in roads, airports, and landfills, to keep layers separate and maintain the structure's strength.

Based on the above brief, the following are the correct and wrong answers. A table can also be found in the *NCEES Handbook version 2.0* page 165 that can be referred to.

☑ **Geotextiles**
Geotextiles act as separators between different layers, preventing them from mixing and maintaining the integrity of each layer and the overall structure.

☑ **Geocomposites**
While geocomposites are primarily used for drainage, they can also serve a separation function. The geotextile layers in geocomposites act as separators, preventing the mixing of different materials, such as soil and aggregate, thereby maintaining the integrity and functionality of both materials.

☐ **Geomembranes**
Geomembranes are used primarily for containments; geomembranes can also serve as a separation layer in some applications.

Further details on geomembranes are provided in Solution 5.14.

☐ **Geogrids**
Geogrids are used for reinforcement and soil stabilization, providing tensile strength to support soil layers.

Further details on geogrids are provided in Solution 5.14.

☐ **Geosynthetic Clay Liner (GCL)**
GCLs are designed for containment purposes, such as lining landfills to prevent leachate migration.

Further details on GCL are provided in Solution 5.14.

☐ **Geocell**
Geocells are used for soil confinement and erosion control, creating a stable base for various applications.

SOLUTION 5.14
The *NCEES Handbook version 2.0*, Section 3.15 Landfills is referred to in this solution.

It is noted that the above referenced section offers a limited selection of options without detailed explanations. Therefore, this solution extends upon the handbook by providing a comprehensive elaboration for each type of geosynthetic utilized, along with the rationale behind their use.

For a more in-depth understanding, feel free to consult with references B.[2] & B.[3] in the bibliography section. If necessary, attempt to obtain them as they offer further insights into these materials.

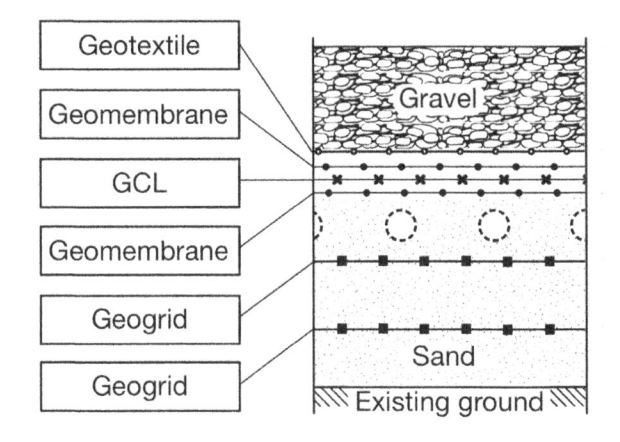

thereby fulfilling a dual purpose of reinforcement and segregation. This is why two layers of geogrid were selected, part of which the ground below is softened from all the disposals as described in the question. The below sketch represents the interlocking power of geogrids:

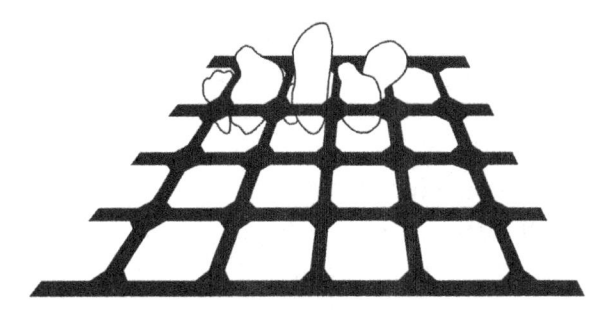

Geosynthetic Clay Liner & Geomembrane Layers
The liner (barrier) system used here employs a composite layer of Geosynthetic Clay Liner sandwiched between two geomembrane layers. This system, or the combination thereof, is widely used. See below further details on this.

Geomembranes are polymer-based sheets that have extremely low permeability to fluids and gases, making them ideal for use as barriers against liquids or vapors. They are often used as foundational or capping liners in landfills as is the case in this question.

GCLs are infiltration barriers consisting of a layer of unhydrated, loose granular or powdered bentonite placed between two or on top of one geosynthetic layer (geotextile or geomembrane). They are highly impermeable when activated and because of this, their function replaces thick layers of clay.

Geotextile Layer
The final geosynthetics layer in this arrangement is a geotextile layer, which

Geogrid Layers
The primary role of geogrids is to provide reinforcement. Their rigidity is significant, and their openings are sufficient to enable interlocking with the adjacent soil or rock,

functions as a protective cushion. It safeguards the underlying composite liner from potential damage caused by the gravel in the drainage layer above it.

SOLUTION 5.15

Despite the absence of any mention of wicks in the NCEES Handbook or the NCEES required list of references on the date of this publication, a thorough understanding of this subject is essential for the exam. A wealth of information on this topic can be found in *FHWA-NHI-16-027 Ground Modification Methods Reference Manual Volume I and II* found in E.[5] & E.[6] in bibliography.

While this publication is not included in the exam list of references now, it is frequently cited in the NCEES Handbook, making it an important reference to familiarize oneself with it prior to the exam.

Information on wicks can be found in Chapter 2 Vertical Drains and Accelerated Consolidate of the above-mentioned reference.

Based on this chapter, it is evident that installing wicks reduces the amount of surcharge or preloading **(making statement I incorrect)**, accelerates the rate of strength gain **(validating statement II as correct)**, but does not completely eliminate post-construction settlement, although it does reduce it **(making statement III incorrect)**.

Furthermore, for this process to be effective, the embankment must be fully saturated before commencement, and it should be normally to slightly over-consolidated **(making statement IV incorrect)**.

Correct Answer is (C)

SOLUTION 5.16

Start with calculating the stress below the three mentioned layers, ignoring the weight of the geofoam as requested in the question:

$$\sigma = (\gamma h)_{asphalt} + (\gamma h)_{base} + (\gamma h)_{subbase}$$

$$= 135\, pcf \times \left(\frac{2}{12}\right) ft$$

$$+125\, pcf \times \left(\frac{10}{12}\right) ft$$

$$+120\, pcf \times \left(\frac{25}{12}\right) ft$$

$$= 376.7\, psf$$

Add to that the traffic load of $200\, psf$ as follows:

$$\sigma_{total} = 376.7 + 200 = 576.7\, psf\, (4.0\, psi)$$

Section 3.13.6 Geofoam of the *NCEES Handbook version 2.0,* is used to choose a geofoam type based on the given strain of 1% and the calculated compressive strength. **The best type in this case is the expanded type EPS19.**

Correct Answer is (D)

SOLUTION 5.17

The *NCEES Handbook version 2.0*, Section 3.12.3 Vibrocompaction, is referred to in this question.

You are encouraged to review the gradation distribution and the categories of soil types suitable for vibrocompaction treatment in this section. Additionally, the sieve analysis information is provided at the top of this section, which can be helpful if you only have a gradation table and not a graph.

The soil gradation ranges have been adjusted based on the provided gradation for scale accuracy. See below:

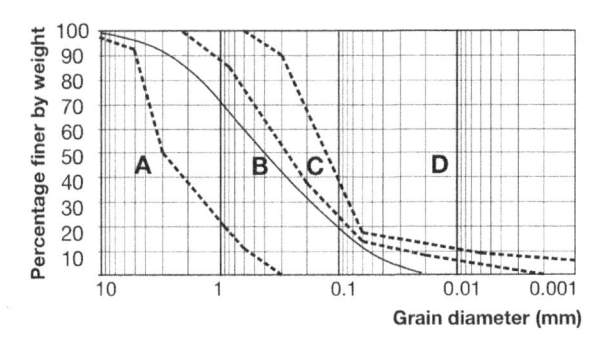

Vibrocompaction Effectiveness:

- **Zone A:** Soils are compactable, but increasing gravel content and high permeability might hinder the depth of penetration. Sand backfills are suggested for these soils during the compaction operation.

- **Zone B: Soils falling in this range are ideal for vibrocompaction with fines content below 10%. Like zone A, a sand backfill is recommended during vibrocompaction.**

- **Zone C:** If the curve falls within this zone, using gravel instead of sand as backfill is advised to enhance vibrator contact and efficiency.

- **Zone D:** Soils with curves in this zone are not easily compacted by vibrocompaction. Stone columns are more suitable for this type of soil.

Stone columns, or vibroreplacement, is a ground improvement method where aggregate columns are created using a water jet to form a hole, then filled with stones. This process compacts the stones and densifies the surrounding soil.

For more information on vibrocompaction, refer to the *FHWA-NHI-16-027 Ground Modification Methods Reference Manual, Volumes I* (reference E.[5] in bibliography), Chapter 4 Deep Compaction. Although not officially listed as an exam reference, it is often mentioned in the NCEES Handbook and is a valuable resource to study before the exam.

Correct Answer is (B)

SOLUTION 5.18

The *NCEES Handbook version 2.0* Section 3.12.1 Types of Ground Improvement has very limited information on Column Supported Embankments CSEs, however, knowledge of this topic is important for the exam. For this, you can find useful information in *FHWA-NHI-16-028 Ground Modification Methods Reference Manual Volumes II,* (reference E.[6] in bibliography) Chapter 6 Column Supported Embankments.

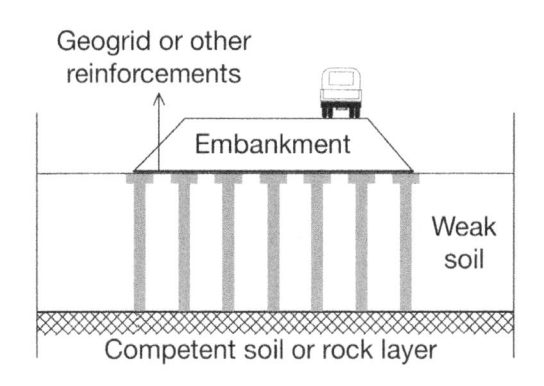

Column Supported Embankments (CSEs) are constructed using stiff vertical columns designed to transfer the load of the embankment through a soft, compressible soil layer to a firm foundation. The primary purpose of a CSE is to transfer the embankment loads through the columns to a competent soil or rock layer beneath the soft foundation soil **(rendering statement I as incorrect)**. See above sketch.

The selection of type of columns used for the CSE depends on several factors, including design loads, constructability, cost, and most importantly, the compressibility of those columns as noted in the *NCEES Handbook version 2.0,* Section 3.12.1 Types of Ground Improvement. If the soil to be supported is weak and cannot bear surface loads, non-compressible columns should be selected **(validating statement II as correct)**.

It is crucial that the load from the embankment is effectively transferred to the columns to prevent the columns from punching through the embankment's Load Transfer Platform LTP, which would cause differential settlement at the surface of the embankment **(rendering statement III as correct)**.

CSEs are typically used to support embankments over soft soil when there is insufficient time for the consolidation of the soft foundation soil using wicks or Prefabricated Vertical Drains (PVDs) and surcharge loads **(making statement IV incorrect)**.

Correct Answer is (B)

SOLUTION 5.19
Reference is made to *FHWA NHI-05-037 Geotechnical Aspects of Pavements,* Section 7.6.5 Admixture Stabilization, and Asphalt Stabilization in pages 7-86 and 7-87.

Based on the above section, the most suitable option is the **Slow Setting Asphalt (SS-1)**. SS-1 is designed to penetrate fine-grained soils and aggregates effectively, providing better stabilization for subgrades with high fines content. It allows for more thorough mixing and bonding with the fine particles, ensuring a more uniform and stable subgrade.

SS-1 is a slow-setting asphalt emulsion. It has a negative charge, which helps it mix well with fine-grained soils and aggregates. This makes it ideal for applications like subgrade stabilization, tack coats, and dust control.

Correct Answer is (D)

SOLUTION 5.20
You can find information on edgedrains in *FHWA NHI-05-037 Geotechnical Aspects of Pavements*, Section 7.2.6 Longitudinal Edgedrains. Additionally, more details are available in publication *FHWA-NHI-16-028 Ground Modification Methods Reference Manual Volumes II,* (reference E.[6] in the bibliography section), Chapter 9 Pavement Support Stabilization Technologies on page 9-104, Section 7.2.1 Construction.

Although the latter reference is not currently listed as a required reference for the exam, it is highly recommended that you obtain a copy of it and familiarize yourself with it due to the wealth of information it contains.

Edgedrains and drain lines should be clearly marked and protected during construction to prevent damage from heavy machinery. However, this method is not foolproof, as operators can still drive over them and cause undetected damage.

A better method is **installing those drains after the pavement construction** to mitigate or avoid this risk entirely. See sketch below:

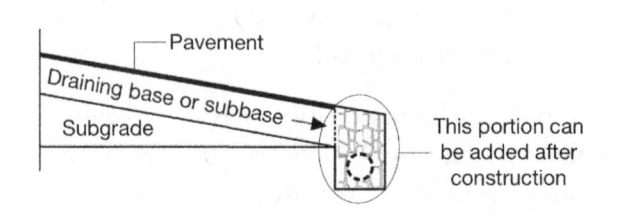

On the other hand, and drawing from the other options provided in this question:

- A concrete rigid pipe is not perforated, which contradicts with the drainage concept and is unsuitable for an edgedrain (making Option A not viable).

- A daylighted permeable layer is irrelevant in this scenario as it is typically used on flat slopes but can also be connected to the edgedrain in other cases (making Option C not applicable).

- Placing the pipe deeper requires verification with the hydraulic model and this may be impractical in most of the cases (making Option D not feasible).

Correct Answer is (B)

PART V
Earth Structures

VI
GROUNDWATER & SEEPAGE

Knowledge Areas Covered

SN	Knowledge Area
6	**Groundwater and Seepage** A. Dewatering, seepage analysis, groundwater flow, and impact on nearby structures B. Drainage design/infiltration and seepage control

PART VI
GW & Seepage

PROBLEM 6.1 *Dewatering Systems*

A contractor has to dewater a 15 ft deep excavation to construct a manhole, the groundwater level in this excavation is nearly 2 ft below ground surface and the type of soil in this excavation is sandy and noncohesive.

Knowing that there are no space requirements around this excavation, the best dewatering system that the contractor can use in this case is the following:

(A) Well points system

(B) Sump pump in the excavation

(C) Deep wells using submersible pumps

(D) Slurry wall

PROBLEM 6.2 *Weep Holes Flow*

The below is a retaining wall cross section with an elevated water table behind it, for which, weep holes were drilled at an interval of 10 ft at its bottom to lower it.

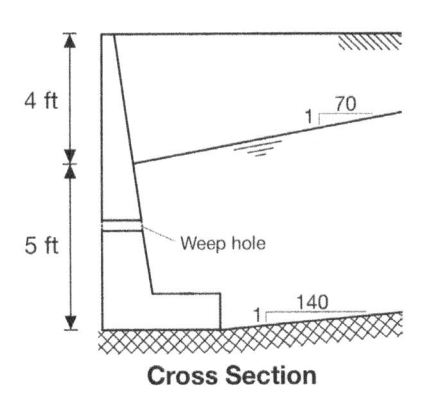

Given that soil permeability (or its hydraulic conductivity) behind this retaining wall is $0.0004\ ft/sec$, the seepage through each weep hole is most nearly:

(A) $5.7\ ft^3/day$

(B) $12.5\ ft^3/day$

(C) $0.5\ ft^3/day$

(D) $25\ ft^3/day$

PROBLEM 6.3 *Dam Seepage (1)*

The below is a cross section of an earth dam that has an isotropic permeability coefficient of $k = 0.0004\ ft/sec$.

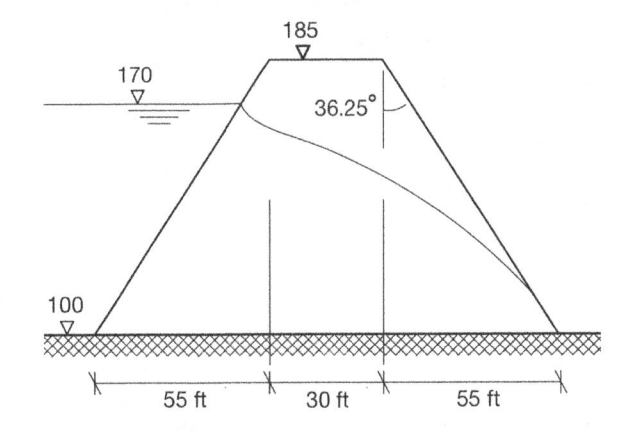

Taking the dimensions shown in the above dam cross section, while knowing that the dam is built on impermeable ground, the expected rate of seepage through this dam is most nearly:

(A) $320 \frac{ft^3}{day}/ft$

(B) $540 \frac{ft^3}{day}/ft$

(C) $750 \frac{ft^3}{day}/ft$

(D) $970 \frac{ft^3}{day}/ft$

PROBLEM 6.4 *Dam Seepage (2)*

The below is a cross section of an earth dam that has an isotropic permeability coefficient of $k = 0.0004\ ft/sec$.

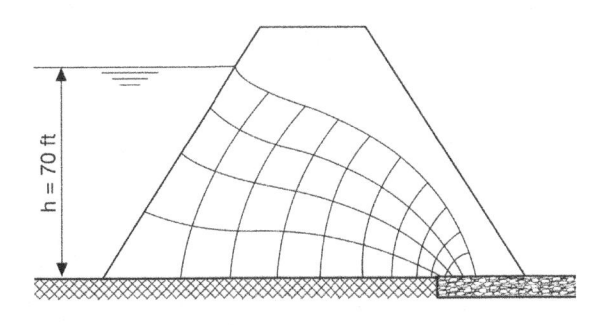

Taking the approximate expression of the flow net as shown in the above dam cross section into account, while knowing that the dam is built on impermeable ground, and provided with a gravel filter as shown, the expected rate of seepage is most nearly:

(A) $320 \frac{ft^3}{day} /ft$

(B) $540 \frac{ft^3}{day} /ft$

(C) $750 \frac{ft^3}{day} /ft$

(D) $970 \frac{ft^3}{day} /ft$

PROBLEM 6.5 *Dam Uplifting Force*

The below is a cross section of a dam that has an isotropic permeability coefficient of $k = 0.0004\ ft/sec$. The dam has a water elevation behind it of $30\ ft$ along with a $3\ ft$ free board.

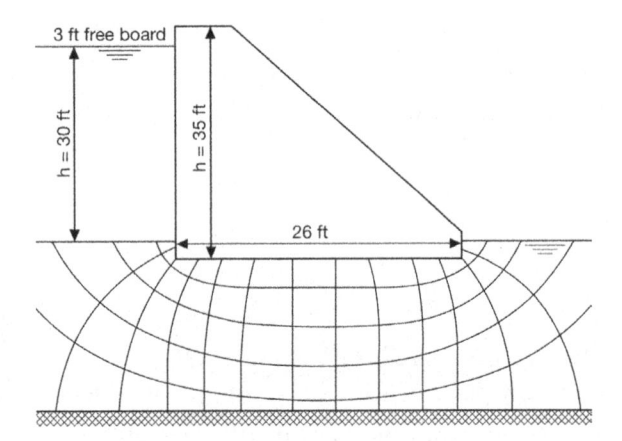

The hydraulic uplifting force that this dam is expected to experience per unit of length of this dam is most nearly:

(A) $15{,}330\ lb\ per\ ft$

(B) $27{,}580\ lb\ per\ ft$

(C) $33{,}450\ lb\ per\ ft$

(D) $44{,}450\ lb\ per\ ft$

🗐 PROBLEM 6.6 *Drainage Design*

Given the excavation site shown in the below sketch (*) where this site is excavated in close proximity to existing buildings:

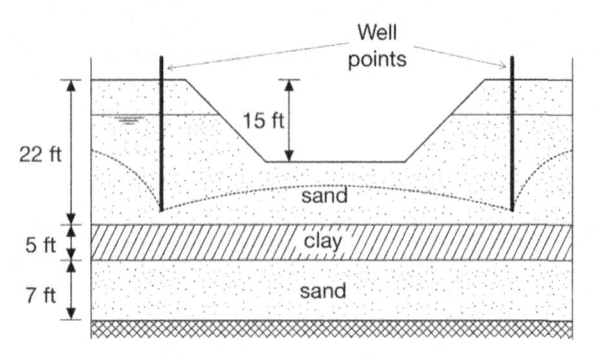

Assuming that the density of the clay layer is $120\ pcf$ and the sand layer is $110\ pcf$, and the depth of groundwater is $5\ ft$ measured from the ground level. The most appropriate method(s) to ensure a safe excavation during dewatering, along with a dewatering protocol that protects the surrounding structures from settlement is (are) as follows:

☐ Install a pressure relief well to manage groundwater pressure.

☐ Increase the excavation depth to improve stability and reduce groundwater pressure.

☐ Use a soil grouting technique to stabilize the soil around the excavation site.

☐ Place a well to recharge groundwater near the buildings to mitigate potential settlement.

☐ Use a high-capacity pump to dewater the excavation site faster and finish construction work faster.

(*) The dotted line in the sketch represents the phreatic line, or the drop in groundwater level due to dewatering from the two wells showing in bold solid lines.

PROBLEM 6.7 *Heaving Safety Factor*

The below is a cross section of a sheet pile operation applied to an excavation with a groundwater level of 4 ft below ground.

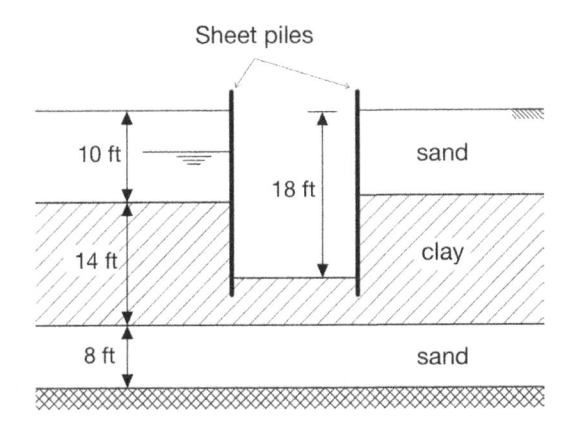

The uplift hydrostatic pressure from the sand layer below the clay layer is $800\ lb/ft^2$, while both the sand and the clay density in this case $= 110\ lb/ft^3$.

The heaving Factor of Safety is most nearly:

(A) 0.8

(B) 1.0

(C) 1.2

(D) 1.4

PROBLEM 6.8 *Cofferdam Operation*

A cofferdam is constructed in a river by driving two sheet piles 22 ft apart and 50 ft long, as shown below.

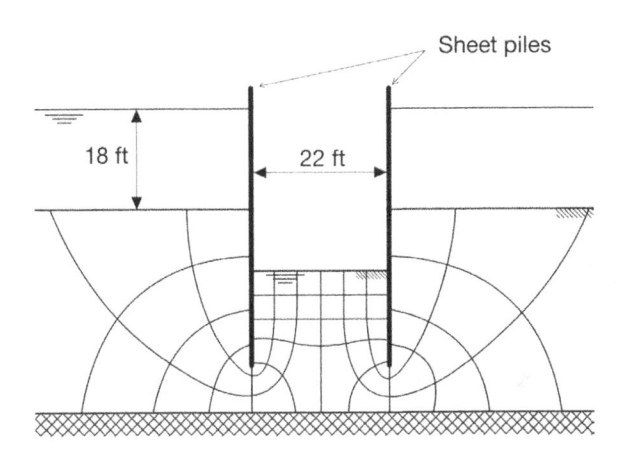

The water depth in the river is 18 ft, and the planned excavation below the riverbed is going to be 10 ft. A flow net has been created, and the soil permeability is isotropic, with a value of $k = 0.0004\ ft/sec$.

The groundwater level needs to be lowered to the level of the excavated riverbed, as indicated on the sketch above. To achieve this, pumps must be installed to continuously draw water from the cofferdam, maintaining the desired water level.

The amount of flow that should be drawn using these pumps in *gallons per minute* (*gpm*) to achieve this criterion is most nearly:

(A) 3,850 gpm

(B) 170 gpm

(C) 85 gpm

(D) 45 gpm

PROBLEM 6.9 *Conventional Well System*

The below is a 40 ft deep excavation with a groundwater level at 1 ft below ground level. The soil encountered in this excavation is sand type with a permeability that allows for a conventional well point system to be installed.

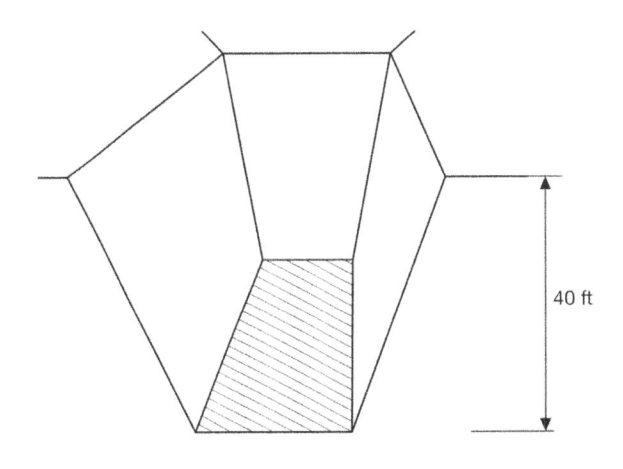

The number of excavation stages for dewatering purposes to be considered for this operation is:

(A) 1

(B) 2

(C) 3

(D) 4

PROBLEM 6.10 *Unconfined Aquifer*

A 100 ft thick unconfined aquifer has a 12 in diameter well that pumps groundwater from it at a rate of 65 gpm (gallon per minute).

Assuming the radius of influence is 450 ft and permeability is $4 \times 10^{-4} ft/sec$, the drawdown at the well is most nearly:

(A) 96 ft

(B) 97.5 ft

(C) 4 ft

(D) 2.5 ft

SOLUTION 6.1

☑ **Wellpoints (*):**

The wellpoint system involves installing small, individual wells around an excavation. These points are connected to a central, centrifugal header pipe with a vacuum function. They effectively lower groundwater levels, creating a stable, dry area for construction works. They are particularly useful in shallow excavations or on job sites with fine-grained soils. Wellpoint dewatering is both cost-effective and easy to install.

☐ **Sump Pumps:**

Sump pumps are more suitable for dewatering non-sandy excavations. They work best for cohesive or hard strata. Although they may work in this scenario with proper trench preparation (such as using geotextile bags), there are better options available.

☐ **Deep Wells:**

Deep wells are specifically designed for deeper excavations. Since the given scenario pertains to a shallow excavation, option C (deep wells) can be ruled out.

☐ **Slurry Walls/Diaphragm Wall Construction:**

A diaphragm wall is constructed to prevent water from seeping into the excavation, especially if the wall extends into a permeable layer. However, diaphragm walls are primarily used for deep excavations, so they are not suitable or noneconomical for this question.

Correct Answer is (A)

(*) You are encouraged to check the *NAVFAC DM – 7.02* Chapter 1, Section 6 Groundwater Control Table 7 for more information on the above.

SOLUTION 6.2

Reference is made to the *FHWA NHI-16-072 Geotechnical Site Characterization,* Section 10.2.2 Water Flow and Hydraulic Conductivity. Equation 10.2 is used as follows (Also found in the *NCEES Handbook version 2.0* Section 3.16.1 Darcy's Law):

$$Q = kiA$$

Where k is the hydraulic conductivity of soil given in the question as $0.0004\ ft/sec$, A is the effective area per weep hole (see below sketch that shows an elevation view for the retaining wall and effective area per weep hole), and i is the hydraulic gradient, or the head over distance for water elevation given as $1:70$.

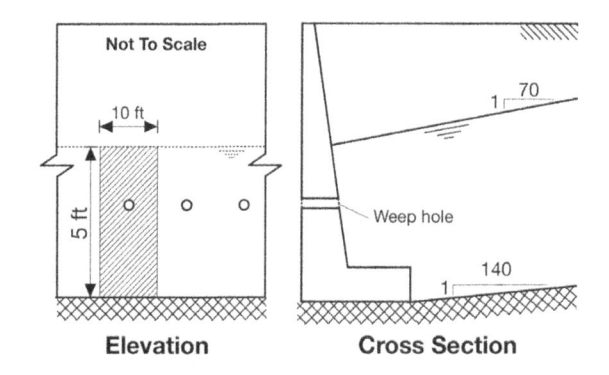

$$Q = 0.0004\ ft/sec \times \frac{1}{70} \times (5 \times 10)\ ft^2$$

$$= 2.86 \times 10^{-4}\ cfs \left(= 24.7 \frac{ft^3}{day}\right)$$

Correct Answer is (D)

SOLUTION 6.3

Reference is made to the *NCEES Handbook version 2.0,* Section 3.14 Earth Dams, Levees, Embankments.

The solution of this question is a direct application to the equation and method presented in page 180 of the handbook. It should be noted however that this method only applies when:

- $\alpha < 60$
- Impervious foundation
- No gravel filter presented
- Isotropic permeability – i.e., horizontal and vertical permeability are equal ($k_h = k_v$)

If any of the above cases do not apply, consult with the cases presented in page 183 of the abovementioned version of the handbook. Or, if given in the question, use the flow net method presented in Solution 6.4 here.

Based on the above references, the following dam dimensions have been constructed using trigonometry:

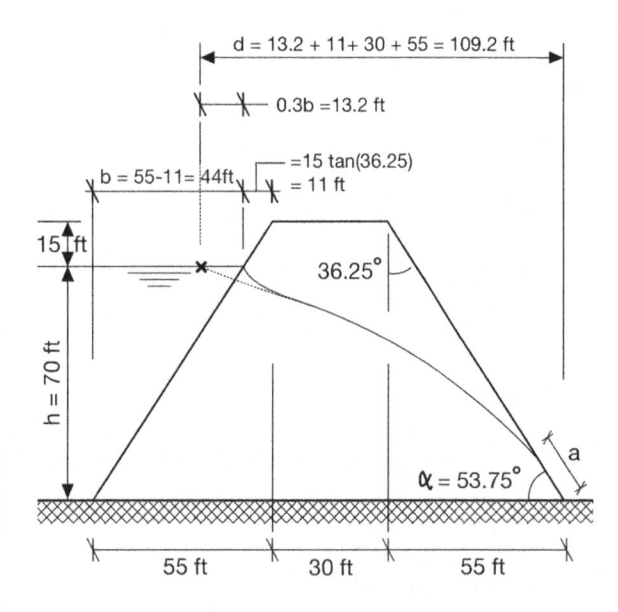

The presented method is used to define the shape and the location of the phreatic line, which is key in estimating the quantity of the flow. The phreatic line is the curved line shown in the above sketch, and it separates the saturated zone, where there is positive hydrostatic pressure, from the unsaturated zone, where there is negative hydrostatic pressure in the dam. See below:

$$S_o = \sqrt{h^2 + d^2}$$
$$= \sqrt{70^2 + 109.2^2}$$
$$= 129.7 \, ft$$

$$a = S_o - \sqrt{S_o^2 - \frac{h^2}{\sin^2 \alpha}}$$
$$= 129.7 - \sqrt{129.7^2 - \frac{70^2}{\sin^2(53.75)}}$$
$$= 33.3 \, ft$$

$$Q = ka \sin^2 \alpha$$
$$= 0.0004 \frac{ft}{sec} \times 33.3 \, ft \times \sin^2(53.75)$$
$$= 8.66 \times 10^{-3} \frac{ft^3}{sec} / ft \left(748.5 \frac{ft^3}{day}/ft\right)$$

Correct Answer is (C)

SOLUTION 6.4

Reference is made to the *NCEES Handbook version 2.0*, Section 3.14 Earth Dams, Levees, Embankments, page 182 Flow Nets.

Flow net is a grid that consists of two set of lines: the flow lines (N_f), and they represent the paths that water particles follow, and the equipotential lines (N_p). The latter connect points within the soil that have the same hydraulic head or potential energy, and they represent drops in head as you move along the flow. The equipotential lines are perpendicular to the flow lines. See below:

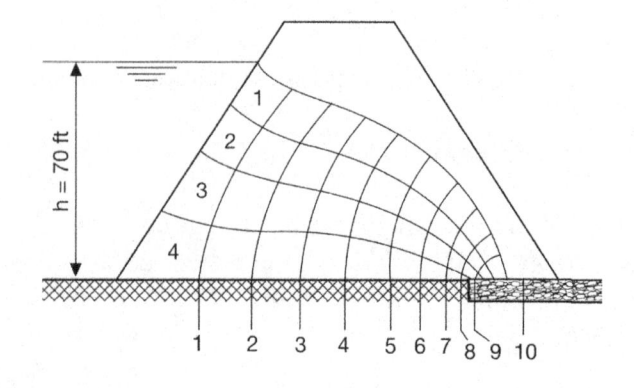

The flow net can be used to estimate the flow as follows:

$$Q = kh\frac{N_f}{N_p}$$

$$= 0.0004\frac{ft}{sec} \times 70\,ft \times \frac{4}{10}$$

$$= 0.0112\,\frac{ft^3}{sec}/ft\,\left(967.7\,\frac{ft^3}{day}/ft\right)$$

Correct Answer is (D)

SOLUTION 6.5

Flow nets can be used to determine the hydraulic uplifting force under a structure. The relevant flow nets have been provided in this question.

The cross section of the dam is replotted below showing the equipotential lines $(N_p) = 13$.

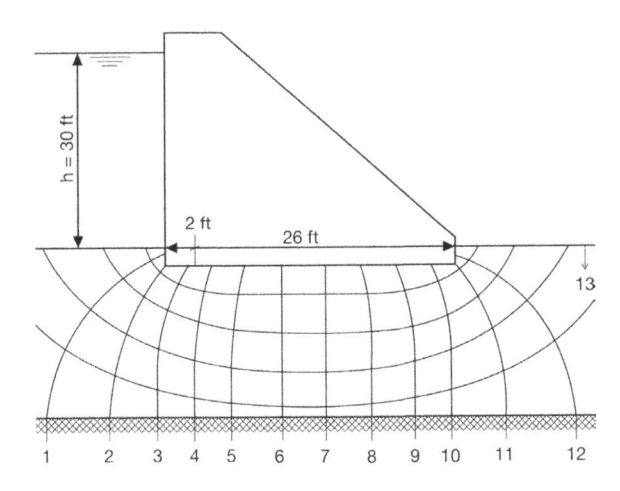

Taking the left bottom edge for this dam, the pressure head at this point equals to $(30\,ft + 2\,ft)$ minus the relevant hydraulic head loss in that location. The hydraulic head loss is calculated using the drops in the equipotential lines (N_p) being two at this location as follows:

$$= 2 \times \frac{h}{N_p} = 2 \times \frac{30}{13} = 4.62\,ft$$

$$P_{left\,edge} = 32 - 4.62 \cong 27.4\,ft$$

The dam's bottom right edge pressure is calculated in a similar fashion as follows:

$$P_{right\,edge} = 32 - 11 \times \frac{30}{13} = 6.6\,ft$$

It can be assumed that the pressure head between the left and right edges – given no variations in the shape or permeability – is linear. See below:

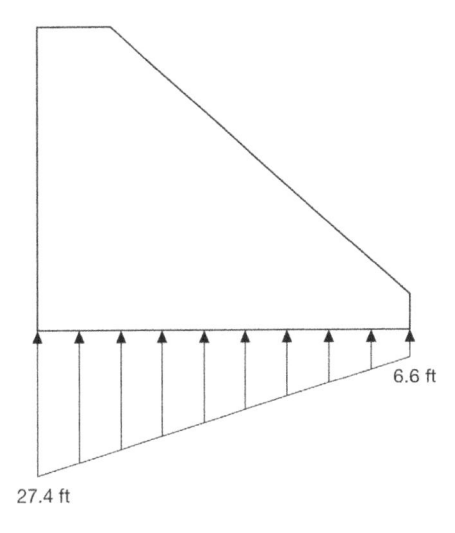

Finally, to calculate the uplifting hydraulic pressure for a $1\,ft$ strip of this dam, multiply the area of the triangle shown above by the density of water as follows:

$$= \left(1\,ft \times 26\,ft \times \left(\frac{27.4 + 6.6}{2}\right)ft\right) \times 62.4\frac{lb}{ft^3}$$

$$= 27{,}580.1\,lb\,per\,ft$$

Correct Answer is (B)

SOLUTION 6.6

Reference is made to the *UFC 3-220-05 Dewatering and Groundwater Control*, Chapter 2 Methods for Dewatering, Pressure Relief, and Seepage Cutoff. Based on this reference, the following are the correct and incorrect answers with explanations provided as needed.

Correct Answers are as follows:

☑ *Install a pressure relief well to manage groundwater pressure.*

Installing a pressure relief well is essential when dealing with artesian wells, and the provided cross section in the body of the question illustrates an artesian flow setup. Artesian conditions involve groundwater under pressure that can cause instability at the bottom of the excavation – See *NCEES handbook version 2.0* Section 3.16.5.3.

The pressure generated by the artesian well at the bottom of the excavation is equal to the water head multiplied by the density of water ($22\,ft \times 62.4\,pcf = 1,373\,psf$), which leads to soil heaving. The remaining soil after excavation provides resistance against heaving ($7\,ft \times 110\,pcf + 5\,ft \times 120\,pcf = 1,370\,psf$). Factor of safety for heaving should be 1.5 to 2 – see UFC 3-220-10 Section 6-5.2.2. See sketch below:

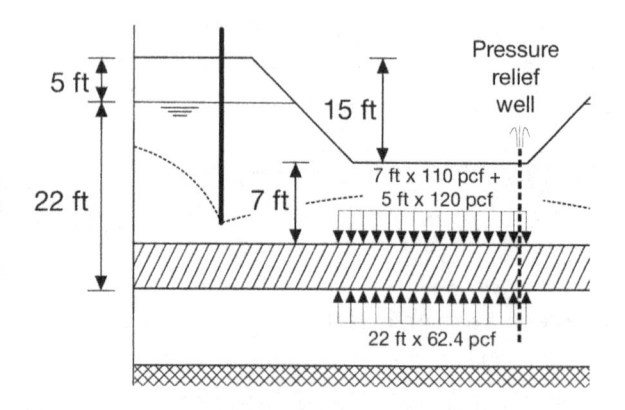

A pressure relief well helps reduce the hydrostatic pressure acting on the excavation walls and floor. This prevents issues such as heaving as discussed in the preceding paragraph, which can lead to instability and potential collapse. By managing this pressure, the overall safety of the excavation is enhanced, and the structural integrity of the excavation site is maintained.

☑ *Place a well to recharge groundwater near the buildings to mitigate potential settlement.*

Placing a well to recharge groundwater near the buildings helps to maintain or restore groundwater levels around the surrounding structures. This method is important as it protects nearby buildings from potential settlement caused by a drop in groundwater levels. By recharging groundwater, the soil is stabilized to prevent differential settlement or subsidence that could damage the nearby structures – See UFC3-220-05 Figure 2.13.

Incorrect Answers are as follows:

☐ *Increase the excavation depth to improve stability and reduce groundwater pressure.*
Increasing the excavation depth does not address groundwater pressure issues. In fact, deeper excavations exacerbate groundwater problems by potentially causing more water to flow into the excavation site. This could lead to higher hydrostatic pressures and greater instability.

☐ *Use a soil grouting technique to stabilize the soil around the excavation site.*
Grouting alone does not compensate for inadequate management of groundwater levels. A comprehensive approach that includes both groundwater control measures (like pressure relief wells) and soil stabilization techniques is necessary for safe excavation practices.

☐ *Use a high-capacity pump to dewater the excavation site faster.*
Rapid dewatering can lead to significant changes in groundwater levels, which can cause instability in the surrounding soil and affect nearby buildings.

SOLUTION 6.7

Reference can be made to Solution 6.6 which explains the concept of heaving, also to *UFC 3-220-10 Soil Mechanics,* Section 6-4.1 Heave.

The below cross section is copied here showing the remaining thickness of the clay layer after excavation, which is the only layer that will provide resistance against heaving.

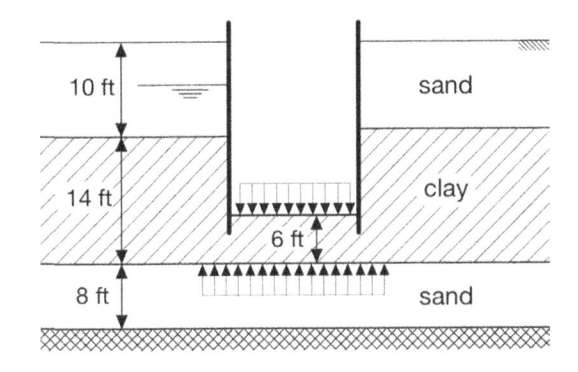

$$FS = \frac{6\,ft \times 110\,lb/ft^3}{800\,lb/ft^2} = 0.825$$

Correct Answer is (A)

SOLUTION 6.8

The concept of net flow is explained in Solutions 6.4 and 6.5. Moreover, reference can be made to the *NCEES Handbook version 2.0,* Section 3.16.4 Flow Nets.

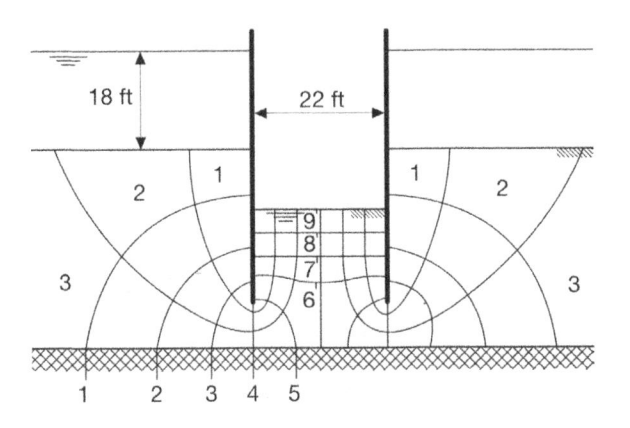

It should be observed in this case that the head considers the excavation depth, and this is already depicted in the equipotential lines provided in the question. See below:

$$Q = kh\frac{N_f}{N_p}$$

$$= \left(0.0004\frac{ft}{sec} \times (18+10)\,ft \times \frac{6}{9}\right) \times 50\,ft$$

$$= 0.373\,\frac{ft^3}{sec}\,(167.4\,gpm)$$

Correct Answer is (B)

SOLUTION 6.9

Reference is made to the *UFC 3-220-05 Dewatering and Groundwater Control,* Section 2.10 Selection of Dewatering System, and Table 2.1 Summary of Ground Control Methods.

It can be noted from the above reference that the maximum drawdown for the conventional wellpoint system is 15 ft only. Based on this, the below proposed stages of excavation can be suggested.

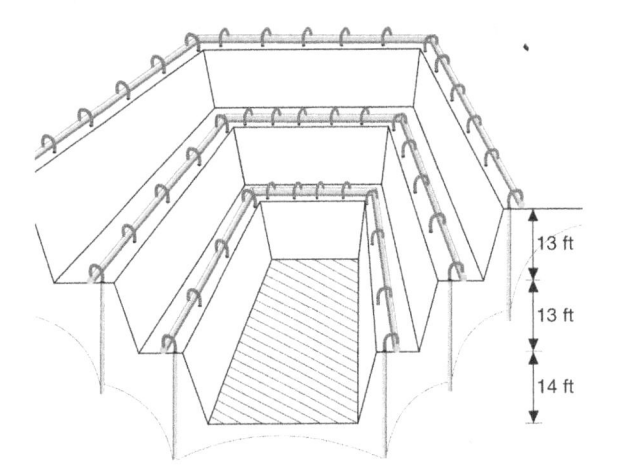

The excavation should be carried out in **three stages**.

Correct Answer is (C)

SOLUTION 6.10
The *NCEES Handbook*, Section 6.6.3.1 (*), Unconfined Aquifers/Dupuit's equation can be used to solve this question.

$$Q = \frac{\pi K(h_2^2 - h_1^2)}{\ln\left(\frac{r_2}{r_1}\right)}$$

Q is the flow rate in ft^3/sec, h_1 and h_2 are heights of the aquifer measured from its bottom at the perimeter of the well (i.e., $r_1 = \frac{12}{2} in = 0.5 \, ft$) and at the influence radius of $r_2 = 450 \, ft$ respectively.

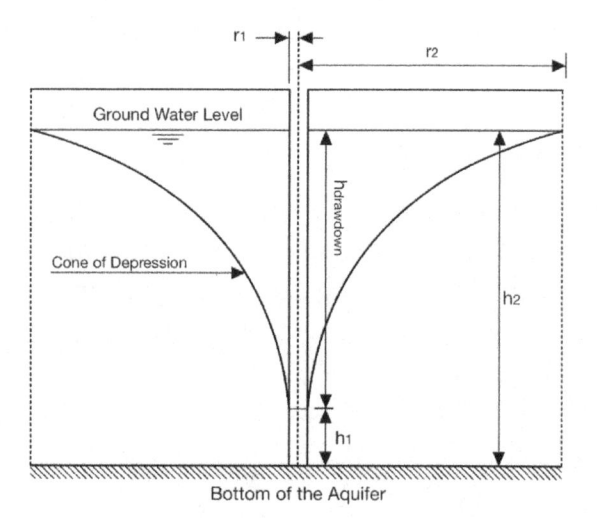

(*) The same equation can be found in other references such as the *FHWA NHI-16-072 Geotechnical Site Characterization*, Section 10.12.8 Field Pumping Tests, page 10-46, Equation 10.23. Attention to be given to numbering as they are different in those two references.

Radius of influence defines the outer radius of the cone of depression, hence $h_2 = 100 \, ft$.

$$h_1 = \sqrt{\frac{\pi K h_2^2 - Q \times \ln\left(\frac{r_2}{r_1}\right)}{\pi K}}$$

$$= \sqrt{\frac{\pi \times 4 \times 10^{-4} \frac{ft}{sec} \times (100 \, ft)^2 - 65 \frac{gal}{min}\left(\frac{0.134 \, ft^3}{gal}\right)\left(\frac{1 \, min}{60 \, sec}\right) \times \ln\left(\frac{450 \, ft}{0.5 \, ft}\right)}{\pi \times 4 \times 10^{-4} \frac{ft}{sec}}}$$

$$= 96 \, ft$$

$h_{drawdown} = 100 \, ft - 96 \, ft = 4 \, ft$

Correct Answer is (C)

VII
PROBLEMATIC SOIL

Knowledge Areas Covered

SN	Knowledge Area
7	**Problematic Soil and Rock Conditions** A. Karst, collapsible, expansive, peat, organic, and sensitive soils B. Reactive/corrosive soils (e.g., identification, protective measures) C. Frost susceptibility D. Rock slopes and rockfalls

PART VII
Problematic Soil

PROBLEM 7.1 *Collapsible Soil*
The following statement(s) are true regarding collapsible soils along with the magnitude of collapse:

I. The magnitude of collapse decreases with the increase in the pre-collapse water content.
II. The magnitude of collapse decreases with the increase of pre-collapse dry density.
III. The magnitude of collapse decreases with the decrease of the pre-collapse overburden pressure.
IV. A negligible amount of collapse occurs when the water content is above a certain percentage.

(A) I + II + III + IV
(B) II + III + IV
(C) I + II + III
(D) I + III + IV

PROBLEM 7.2 *Identifying Problematic Soils*
Soil samples were collected from a certain field location with potential problematic soils, the natural dry density for the soil samples was found to be $60\,pcf$ and the liquid limit was 50%.

Based on the above information, what could possibly be the characteristics of the problematic soil that has been identified on this field:

(A) Expansive soils
(B) Shrinking soils
(C) Collapsible soils
(D) Liquifiable soils

PROBLEM 7.3 *Soil Sensitivity*
Soil swell sensitivity is mainly determined by the following:

(A) Whether the clayey soils include kaolinite.
(B) Whether the clayey soils include smectite.
(C) Whether soil is disturbed or not.
(D) The depth the soil is found at.

PROBLEM 7.4 *Identifying Organic Soils*
The most effective test that can be used to identify and/or classify organic soils is the following:

(A) Cone Penetration Test (CPT)
(B) Dilatometer Test (DMT)
(C) Chemical tests
(D) Determine ash content and water content

PROBLEM 7.5 *Identifying Dispersive Soils*
The following test can be used to identify dispersive soils using indirect measurements:

(A) Chemical tests
(B) Double hydrometer test
(C) Cone penetration test
(D) Pinhole dispersion test

PROBLEM 7.6 *Soil Liquefaction*
The following statement(s) represent some of the key qualifications for soil liquefaction to occur:

I. Liquefaction can occur to any type of soil when saturated with water while combined with any magnitude of earth shaking.

II. Soil composition is coarse grained with less than 20% fines.
III. Soil below water table is more prone to liquefaction when significant earth excitement occurs.
IV. Relative density when above 60% but less than 90% could be a major contributing factor to liquefaction upon earth shaking.

(A) I + II + III + IV
(B) II + III + IV
(C) II + III
(D) III + IV

PROBLEM 7.7 *Corrosive Soils*
The following statement(s) represent some of the characteristics of soils that have high corrosion potential:

I. Soils fully submerged in water are highly likely to have higher corrosion rates.
II. Soils with high resistivity are more prone to become corrosive compared to soils with low resistivity.
III. Soils with a low $pH < 4$ (extremely acidic) are more prone to being corrosive compared to soils with $pH > 10$ (extremely alkaline).
IV. The higher the concentration of soluble salts in soils the more potential of it to become corrosive.

(A) I + II
(B) II + III
(C) III + IV
(D) IV

PROBLEM 7.8 *Corrosivity Scoring*
The corrosivity score for a soil located at a certain construction site, which is moist most of the time, with fair drainage provided, a *pH* value of 7 and resistivity of 4,000 $\Omega \cdot$ cm is most nearly:

(A) 2
(B) 5
(C) 9
(D) 14

PROBLEM 7.9 *Sensitive Clay*
Sensitive clay is identified with the use of the following index:

(A) Liquidity index
(B) Plasticity index
(C) Shrinkage limit
(D) Plastic limit

PROBLEM 7.10 *Exploratory Drilling*
A 10 ft diameter circular foundation is to be constructed 8 ft below the ground surface.

This foundation is located in a karst environment, and exploratory holes are to be drilled at the location of this foundation to make sure there are no cavities below it and to confirm the existence of competent material.

The depth of the exploratory drilling holes in this case should extend well beyond _____ ft from the surface of earth:

(A) 18 ft
(B) 23 ft
(C) 28 ft
(D) 56 ft

PROBLEM 7.11 *Sink Holes*

The below is the main reason why sinkholes form and cause disturbance especially when located at a major road arterial:

(A) The dissolution of dolomite and limestone.

(B) Earthquakes and seismic activities at lose ground.

(C) Areas with heavy rain where the base material is poorly graded or loose.

(D) Leakage from a broken pipe.

PROBLEM 7.12 *Frost Susceptible Soils*

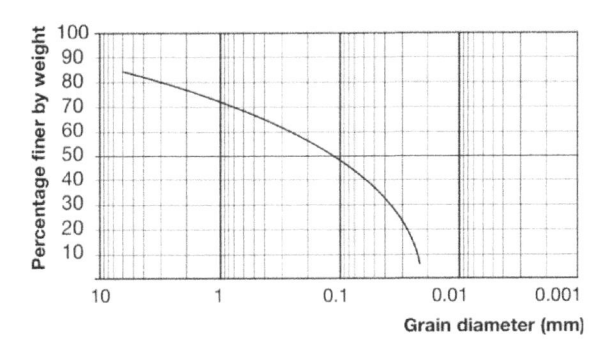

The above gradation is for a location covered with silty sand with a depth of at least 10 ft. You have been asked to assess the susceptibility of this soil to frost for further treatment if needed.

Based on the above gradation, the group, or the degree of frost susceptibility for this location is:

(A) F1 – Negligible to low

(B) F2 – Low to medium

(C) F3 – High

(D) F4 – Very high

PROBLEM 7.13 *Subgrade Problem Soil*

Upon investigating a subgrade layer, the designer found the that the resilience modulus is low due to permanent high groundwater level.

Following are some measures that could be implemented to improve this situation:

I. Remove water-soaked and weak soil and replace it with a new soil.
II. Place some separator layers that have high voids.
III. Increase the thickness of the base layers.
IV. Place a thick embankment for a period of time prior to construction to expel excess water.
V. Use geogrid to strengthen unbound layers.

(A) I + V

(B) II + III + IV

(C) III + IV + V

(D) II + III + V

PROBLEM 7.14 *Frost Susceptible Soil*

In order to economically minimize the effect of frost penetration into soils with a deep and thick frost heave potential soil layer, one to of the below measures can be used:

(A) Remove and replace all this layer with non-frost susceptible material.

(B) Place an additional non-frost susceptible material to increase depth to frost susceptible soil.

(C) Include drains in this layer to reduce saturation and prevent freezing altogether.

(D) Use geogrid or other material to strengthen this layer when it gets frozen.

PROBLEM 7.15 *Rock Slope Stability*

The below are different patterns of rock slopes with discontinuities. The rock slope that has a higher potential of degrading relatively more quickly if water seeps into its discontinuities from surface runoff while increasing its rate of weathering and potential failure is:

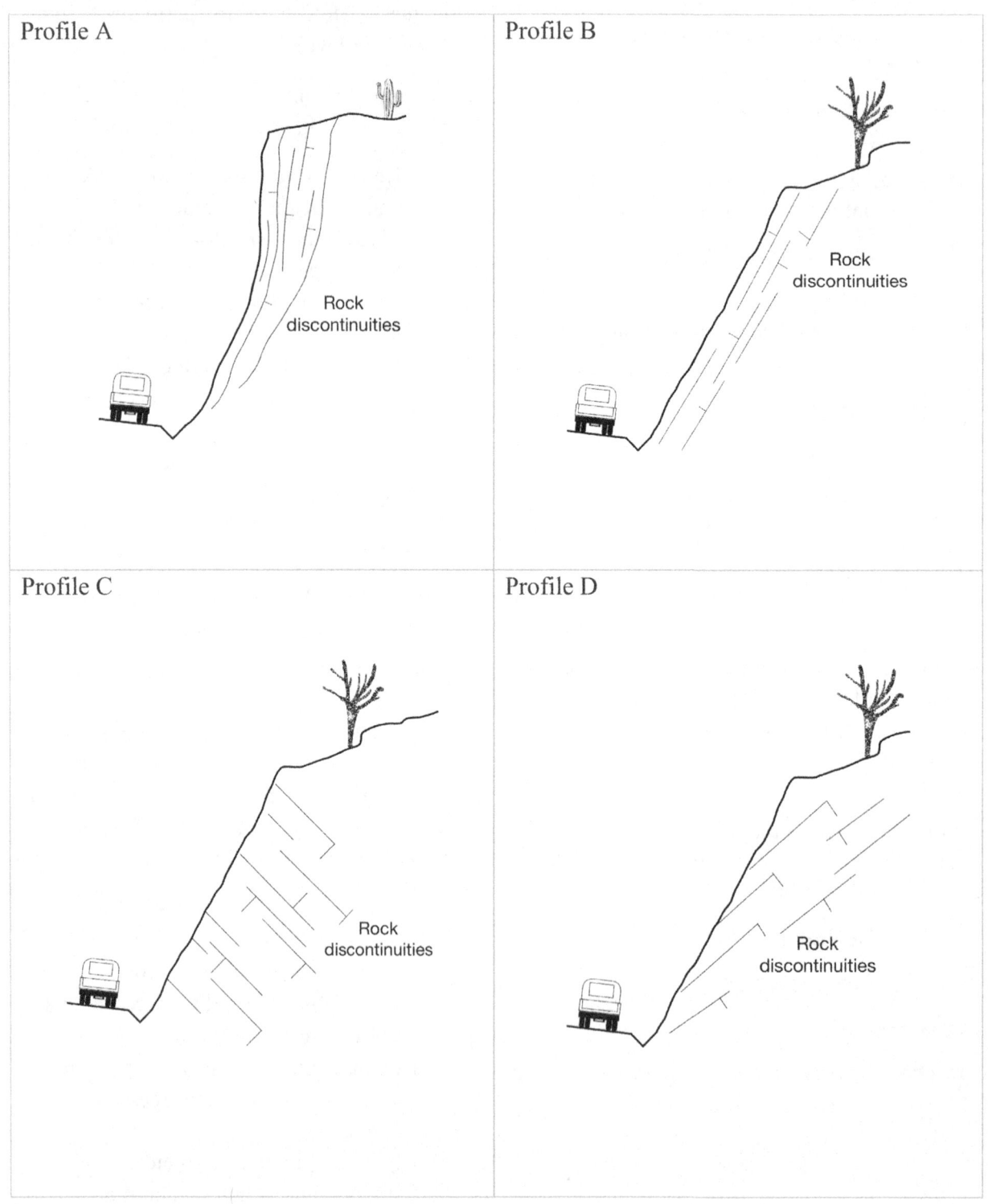

Profile A

Profile B

Profile C

Profile D

SOLUTION 7.1

Reference is made to the *FHWA NHI-16-072 Geotechnical Site Characterization,* Section 5.2 Collapsible Soils, page 5-5.

The last paragraph of this section explains that the magnitude of collapse depends on numerous factors, part of which are mentioned in the statements of this question. Although answers can be drawn from this paragraph alone, the following text shall provide an explanation for each of those cases.

The magnitude of collapse decreases with the increase in the pre-collapse water content.
Capillary forces are stronger with higher initial water content. These forces hold the soil particles together and resist collapse. This means that when soils are more saturated prior to the collapse, these capillary forces could prevent the collapse or reduce its magnitude. Moreover, with more water content, air entry decreases, which prevents water reentry and the change of soil composition which could lead into a higher collapse magnitude.

This marks statement I and IV as correct.

The magnitude of collapse decreases with the increase of pre-collapse dry density.
Higher pre-collapse dry density generally implies better particle packing within the soil structure. Well-packed soils have fewer voids and more interparticle contact and interparticle bonding leading to reduced magnitude of collapse.

This marks statement II as correct.

The magnitude of collapse decreases with the decrease of the pre-collapse overburden pressure.
Although an increased overburden pressure can increase the effective stress, which in turn increases the bond between soil particles and could potentially lead to preventing the collapse, however when a collapse occurs due to other reasons, its magnitude will be higher compared with a lower overburden pressure.

This marks statement III as correct.

Correct Answer is (A)

SOLUTION 7.2

Reference is made to the *FHWA NHI-16-072 Geotechnical Site Characterization,* Section 5.2.2 Indirect Identification of Collapsible Soils, page 5-6.

The below figure is copied from the above reference and is used by Gibbs & Bara (1967) to provide an initial (inconclusive) assessment for problematic soils.

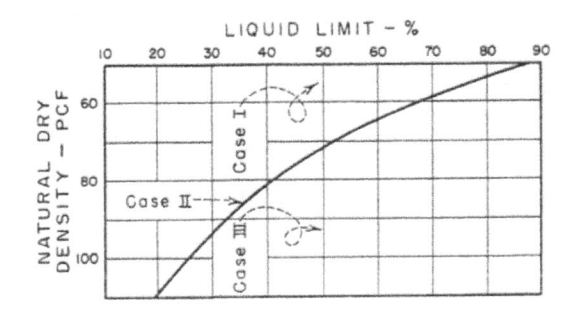

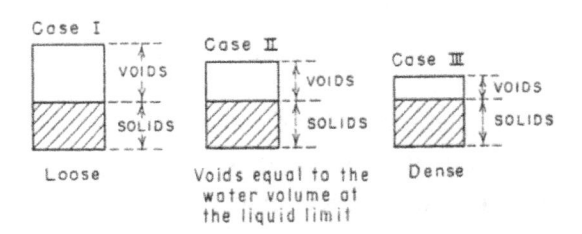

Per Gibbs & Bara (1967), when the dry density along with the liquid limit plot above the shown curved line, **this indicates a possible collapsible soil**. Further tests would be required for direct confirmation in the lab.

Correct Answer is (C)

SOLUTION 7.3
Reference is made to the *FHWA NHI-16-072 Geotechnical Site Characterization,* Section 5.3.4 Swell "Sensitivity".

As the section above indicates, **remodeling / disturbing soils affects their shrink/swell behavior by destroying the natural bonds between soil particles**. These bonds are what protect the soil from swelling in the first place. Swell sensitivity in this case serves as an indication of the magnitude of soil swell when disturbed.

On the other hand, the presence of either kaolinite or smectite determines the soil's composition from the outset and influences the magnitude of swelling. Specifically, kaolinite results in less swelling compared to the presence of smectite. For further details, you can refer to page 5-11 of the same reference under Section 5.3.

Correct Answer is (C)

SOLUTION 7.4
Reference is made to the *FHWA NHI-16-072 Geotechnical Site Characterization,* Section 5.4.2 Identification of Organic Soils.

As mentioned in the referenced section, the most effective method for identifying and understanding the proportions and classification of organic soils is through ASTM methods. **These methods involve measuring the ash content, in situ water content, and the specific gravity of the soil.**

In situ tests such as Cone Penetration Test (CPT), Dilatometer Test (DMT), and others can provide an indication of whether the encountered soil can be classified as organic. For instance, a low tip resistance and high friction ratios in the CPT test may suggest organic content. However, it is important to note that clayey soils and organic soils can exhibit similar physical behaviors in the field. As a result, these methods are considered less effective in accurately identifying organic soils.

Correct Answer is (D)

SOLUTION 7.5
Reference is made to the *FHWA NHI-16-072 Geotechnical Site Characterization,* Section 5.5.2 Identification of Dispersive Soils Using Indirect Methods.

Dispersive soils are problematic, and they are unique in their composition because they can break down and erode rapidly when exposed to water. These soils are prone to erosion, which can lead to tunnel and gully formation. The main reason behind this is its particles expulsion.

For example, if used as a base material for roads or embankments, they can lead to subsidence, instability, and road failure. They can affect building foundations as well.

Sodium ions in the soil is the main reason and the main chemical attribute that causes this behavior. The quantity of exchangeable sodium can be quantified using **chemical tests**. This method however does not constitute a perfect indicator and hence it is only used to indirectly measure or identify dispersive soils.

Correct Answer is (A)

(*) You are encouraged to read about the direct methods for measuring dispersion (i.e., the pinhole dispersion test and the double hydrometer test) of Section 5.5.2 pages 5-28 and 5.29 of the same reference.

SOLUTION 7.6
Reference is made to the *FHWA NHI-16-072 Geotechnical Site Characterization,* Section 5.6.1 Occurrence of Liquefiable Soils.

In the abovementioned section, page 5-31, explains that liquefaction is dependent on ground characterization and the magnitude of shaking. In which case, the key conditions that only apply in this question are when soils are submerged in water and significant shaking occurs **(statement II)**, and, when fines are less than 20% **(statement III)**.

Correct Answer is (C)

SOLUTION 7.7
Reference is made to the *FHWA NHI-16-072 Geotechnical Site Characterization,* Section 5.9.2 Identification of Corrosive Soils.

Soils fully submerged in water are highly likely to have higher corrosion rates.
Soils in the fluctuation zone of groundwater are the ones more susceptible to corrosion due to the presence of moisture and oxygen. Conversely, when soils are fully submerged, they lack access to oxygen, and if there was a continuous flow of groundwater, the latter helps clean the soil particles.

This marks Statement I as incorrect.

Soils with high resistivity are more prone to become corrosive compared to soils with low resistivity.
Resistivity is a broader term that indicates soil corrosivity where high resistivity slows corrosion reactions. For example, sandy soils with resistivity $> 20,000 \, \Omega \cdot cm$ are the least corrosive, while clay soils with around $1,000 \, \Omega \cdot cm$ are highly corrosive.

This marks Statement II as incorrect.

Soils with a low pH < 4 (extremely acidic) are more prone to being corrosive compared to soils with pH > 10 (extremely alkaline).
The two extremes (i.e., extreme acidity and extreme alkalinity) both have a significant potential for higher corrosion rate, but not one over the other.

This marks Statement III as incorrect.

The higher the concentration of soluble salts in soils the more potential of it to become corrosive.
Soluble salts represented with chlorides and sulfates in the soil are directly proportional to corrosivity.

This marks statement IV as correct.

Correct Answer is (D)

SOLUTION 7.8
Reference is made to the *FHWA NHI-16-072 Geotechnical Site Characterization,* Section 5.9.3 Classification of Corrosive Soils, and Table 5.15 Typical Numerical Corrosivity Scoring System.

Points and numerical values from the abovementioned table are assigned for the described condition and then summed up to form a score as follows:

For resistivity of $4,000 \, \Omega \cdot cm$, a score of 2 is assigned, for *pH* of 7, a score of 6 is assigned, and for moisture, a score of 1 is assigned.

$$Score = 2 + 6 + 1 = 9$$

Table 5-16 of the same section indicates that this soil is an aggressive soil when it comes to corrosion potential.

Correct Answer is (C)

SOLUTION 7.9
Reference is made to the *FHWA NHI-16-072 Geotechnical Site Characterization,* Section 5.11.2 Identification of "Quick" Clays.

Sensitive clay, also known as quick clay or quicksand, is a type of clay that can undergo rapid and dramatic change in its state when disturbed. These clays are normally stable, however, when subjected to sudden loads, or vibration or any type of disturbance (e.g., seismic activity), they lose their strength and behave as if they were fluid.

The **liquidity index** is a reliable measure for identifying quicksand. As the liquidity index increases, the sensitivity of the soil also increases. The above reference mentions that highly sensitive clays tend to have a liquidity index > 1.5 or 2.

Correct Answer is (A)

SOLUTION 7.10
Reference is made to the *FHWA NHI-16-072 Geotechnical Site Characterization,* Section 12.2.2 Identification and Characterization of Karst Hazards, and specifically page 12-7.

In this case, the geotechnical engineer should observe the elastic stress distribution below the circular foundation.

From the stress profile presented in the abovementioned reference, it shows that at a depth of two diameters the stress drops to less than 10% of the foundation distributed load. This is located at $8 + 2 \times 10 = 28\ ft$ from ground level.

The reference recommendation in this case is to extend the exploratory holes' depth well beyond $28\ ft$ to confirm the existence of competent material.

Correct Answer is (C)

SOLUTION 7.11
Reference is made to the *FHWA NHI-16-072 Geotechnical Site Characterization,* Section 12.2 Karst Hazards.

While all the options could contribute to the formation of sinkholes when combined with other factors, the significant and main factor is the **dissolution of soluble bedrock, mainly dolomite and limestone**. Over time, as water gradually dissolves and carries away the rock beneath the surface, a cavity or a void can develop. When the overlying sediment or soil can no longer support its own weight, it collapses into a void forming a sinkhole.

Correct Answer is (A)

SOLUTION 7.12
Reference is made to *FHWA NHI-05-037 Geotechnical Aspects of Pavements,* Section 7.5.6 Frost-Susceptible Soils and Table 7-12. Frost susceptibility classification of soils from the same section.

Per the gradation provided in this problem, it can be noticed that that percentage finer than $No.\ 200$ (i.e., $0.075\ mm$) is 45% which puts this soil at the **F4 group (very high susceptibility)**. (*)

You can check page 7-62 and 7-63 for possible treatments for different groups of frost susceptible soils.

Correct Answer is (D)

(*) Frost-susceptible soils are those with pore sizes that promote capillary flow. Like silty and loamy types, are prone to frost heaving due to their fine particles and capillary flow. In contrast, well-drained soils are less susceptible because they retain less water.

SOLUTION 7.13

Reference is made to *FHWA NHI-05-037 Geotechnical Aspects of Pavements,* the following sections are referred to:

- o Section 6.4 Subgrade Stabilization
- o Section 7.2.10 Separator Layers
- o Section 7.3.4 Base Reinforcement

Removing water-soaked soil may not be practical due to cost and time constraints. Instead, consider adding well-draining layers with a high void ratio to prevent water accumulation and further weaken the subgrade. Increasing the depths of subbase and base layers can enhance their structural strength (SN). Lastly, incorporating a geogrid – which is a type of geosynthetics – helps stabilize the affected layers (see Solution 5.13 for some details on the various types of geosynthetics and their use).

This makes statements II, III and V true.

Correct Answer is (D)

SOLUTION 7.14

The most effective and cost-efficient method to address the issue of a thick frost-susceptible layer is to **augment it with an additional layer**, thereby enhancing the depth to the frost line as suggest in Option (B).

Option (A), which involves removing the entire layer, is not a financially viable solution. Option (C), the installation of drainage systems, fails to rectify the fundamental structure and problematic characteristics of the soil. Lastly, the use of geogrids as per Option (D) is not applicable in this scenario.

Correct Answer is (B)

SOLUTION 7.15

For a comprehensive understanding of discontinuities, consult with *UFC-3-220-10 Soil Mechanics*, Section 7-10.1 Modes of Rock Slope Failure, which provides detailed insights into the types of discontinuities and methods of protection. Additionally, *FHWA NHI-16-072*, Section 12.6.1 Implications of Landslide and Rockfall Hazards for Transportation Projects, and *FHWA NHI-11-032* Section 7.4.3 Rock Slopes are both highly recommended for review.

Profile A shows a series of discontinuities, taking the form of columns, that are susceptible to rapid deterioration. The infiltration of water into these discontinuities, and behind those columns, could cause rapid deterioration and reduces slope stability. Moreover, deterioration could be exacerbated if followed by freezing temperatures as this could lead to the expansion and subsequent extrusion of those columns. This latter may result in a shift of the structure's center of gravity, worsening the degradation and potentially leading to a collapse – see sketch below.

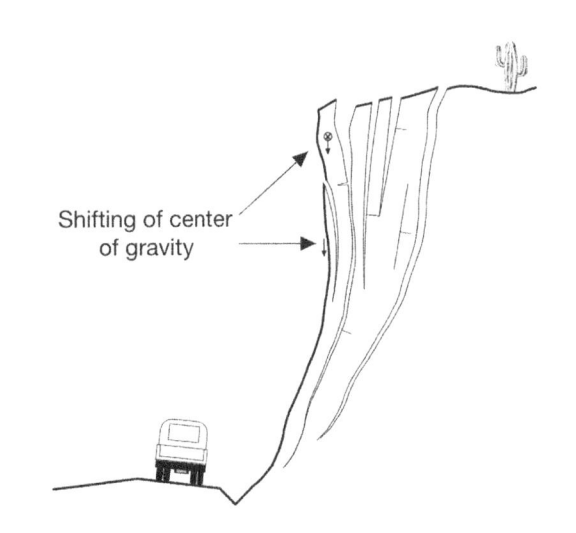

Correct Answer is (A)

PART VII
Problematic Soil

VIII
WALLS & FOUNDATIONS

Knowledge Areas Covered

SN	Knowledge Area
8	**Retaining Structures (ASD or LRFD)** A. Lateral earth pressure and load distribution B. Rigid retaining wall analysis (e.g., CIP, gravity, external stability of MSE, soil nail, crib, bin) C. Cantilevered, anchored, and braced retaining wall analysis (e.g., soldier pile and lagging, sheet pile, secant pile, tangent pile, diaphragm walls, temporary support of excavation, and beams and column elements) D. Cofferdams E. Underpinning methods and effects on adjacent infrastructure F. Ground anchors, tie-backs, soil nails, and rock anchors (e.g., design and quality control)
9	**Shallow Foundations (ASD or LRFD)** A. Bearing capacity B. Settlement, including induced stress distribution
10	**Deep Foundations (ASD or LRFD)** A. Geotechnical and structural capacity and settlement of deep foundations (e.g., driven pile, drilled shaft, micropile, helical screw piles, auger cast piles, beam/column) B. Lateral capacity and deformation of deep foundations C. Installation methods D. Static and dynamic load testing E. Integrity testing methods

PART VIII
Walls & Foundations

PROBLEM 8.1 *Retaining Wall Safety Factors*

The figure below is for a plain concrete gravity retaining wall with an 8 ft Clayey Silt (*) backfill and base.

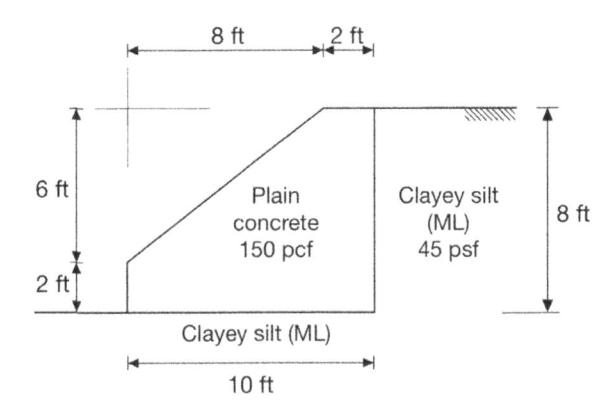

Factors of Safety against sliding and overturning for this wall are as follows:

(A) 0.9 for sliding and 13.2 for overturning.

(B) 1.1 for sliding and 1.3 for overturning.

(C) 2.9 for sliding and 13.2 for overturning.

(D) More information required.

(*) Cohesion exhibited by clayey silts is 130 psf.

PROBLEM 8.2 *Combined footing Dimensions*

The below is a combined footing with the following column and soil properties:

- Column 1 dimensions: 24 in × 12 in
- Column 2 dimensions: 24 in × 24 in
- Service Load $P1$ = 50 kip
- Service Load $P2$ = 75 kip
- Allowable soil pressure = 2 ksf

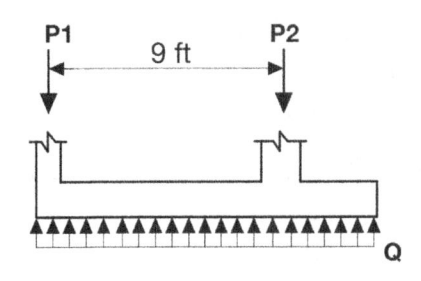

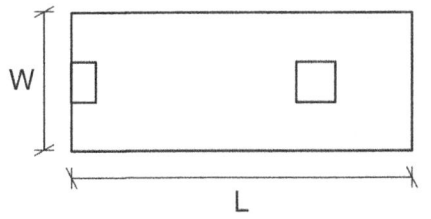

In order to maintain a uniform soil pressure underneath the footing, the following dimensions $W \times L$ should be used:

(A) 6.0 ft × 10.5 ft

(B) 5.5 ft × 11.3 ft

(C) 5.3 ft × 11.8 ft

(D) 5.1 ft × 12.3 ft

PROBLEM 8.3 *Distribution of Pressure under Footing*

The below is a plan view for a single concrete footing with a 12 in × 24 in column that sits directly at a property line carrying the following loads:

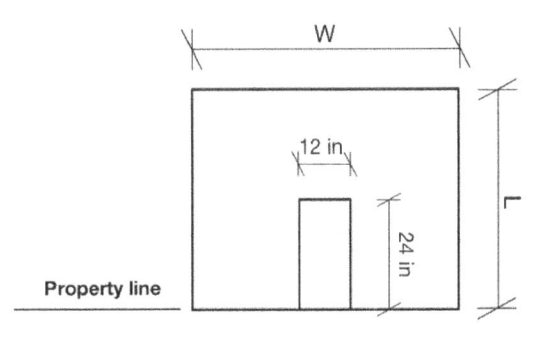

Load type	Service load (kip)	Ultimate load (kip)
Live load	16	19.2
Dead load	11	13.2
Total load	27	32.4

The substrate below this footing is an improved sandy gravel with a load-bearing allowable pressure of 3,000 psf.

The best dimensions $W \times L$ for this footing that generate a safe with no negative pressure zones is:

(A) 8.0 ft × 3.0 ft

(B) 6.0 ft × 3.0 ft

(C) 4.3 ft × 4.3 ft

(D) 3.0 ft × 8.0 ft

PROBLEM 8.4 *Pressure Under Footing*

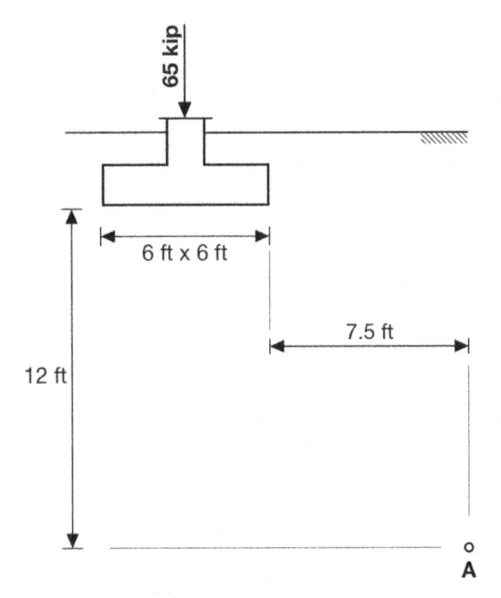

The increase in the vertical pressure at point 'A' due to loading the square footing using those two theories respectively:

o The Boussinesq's Theory

o The 2:1 theory

are as follows:

(A) 110 psf (Boussinesq) and zero (2:1 theory)

(B) 110 psf (Boussinesq) and 200 psf (2:1 theory)

(C) 270 psf (Boussinesq) and zero (2:1 theory)

(D) 270 psf (Boussinesq) and 200 psf (2:1 theory)

PROBLEM 8.5 *Foundation Settlement*

The below is a 6 ft × 6 ft square concrete footing laid in fine medium dense sand with a maximum load applied on it of 100 kip.

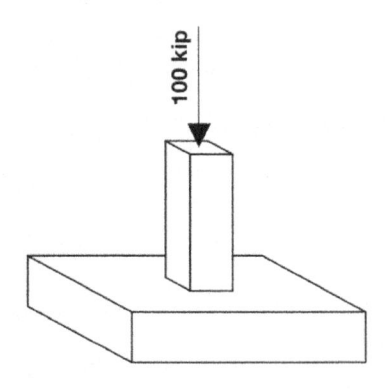

The maximum initial elastic vertical settlement this footing will experience is most nearly:

(A) 0.77 in

(B) 0.064 in

(C) 1.55 in

(D) 0.46 in

PROBLEM 8.6 *Bearing Capacity for a Square Foundation*

The below is a concrete square foundation placed 5 ft below surface in soil that has the following properties:

o Cohesion = 450 psf
o Friction angle = 35°
o Density = 130 pcf
o Groundwater 5 ft deep

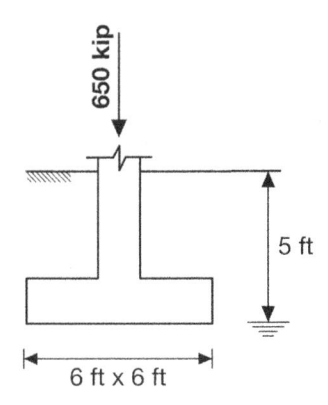

Considering the single footing carries a load of 650 kip, safety factor for this footing against shear failure is most nearly:

(A) 4.3

(B) 2.9

(C) 3.6

(D) 4.6

PROBLEM 8.7 *Construction Operation Over a Single Footing*

During an operation of removing soil from over an embedded foundation for an existing building, the following can happen:

(A) Stress relief as the load which was exerted by the soil on top is removed.

(B) Bearing capacity reduction with possible shear failure.

(C) Reduction in stresses at the bottom reinforcements of the footing.

(D) None of the above.

(*) PROBLEM 8.8 *Retaining Wall Safety Factor*

The below reinforced concrete cantilever retaining wall has the following properties:

- Concrete density $\gamma_{concrete} = 150\ pcf$
- Soil density $\gamma_{soil} = 130\ pcf$
- Water density $\gamma_{water} = 62.4\ pcf$
- Soil's friction angle $\emptyset'_{soil} = 37°$
- Groundwater level 6 ft below surface

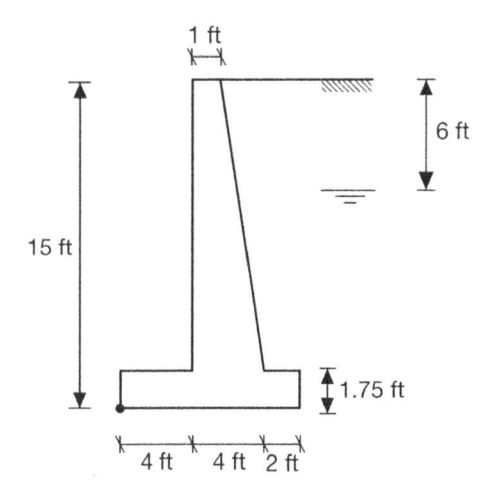

Based on the above information, wall overturning Safety Factor (*) is most nearly:

(A) 2.1

(B) 2.7

(C) 3.7

(D) 15.9

(*) Consider overturning will occur around left most bottom concrete edge.

(✱) PROBLEM 8.9 Ground Anchor Capacity

The below reinforced concrete cantilever retaining wall has the following properties:

- Concrete density $\gamma_{concrete} = 150\ pcf$
- Soil density $\gamma_{soil} = 130\ pcf$
- Soil's friction angle $\phi'_{soil} = 35°$
- Anchor(s) shown below placed at 3 ft intervals along the length of the wall.

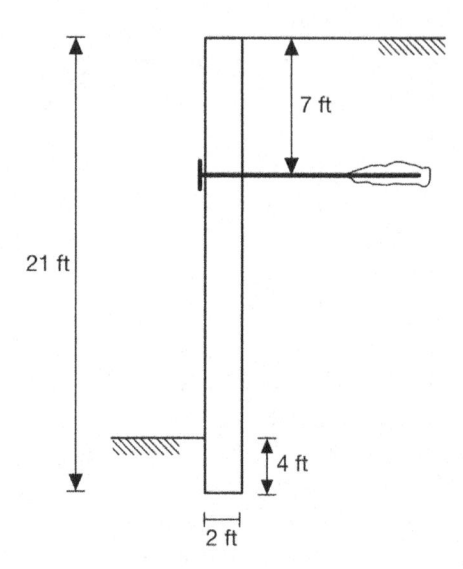

The expected anchor loading using the tributary area method is most nearly (*):

(A) 8 kip

(B) 2 kip

(C) 24 kip

(D) 16 kip

(*) Use $p = k_a\ H\ \gamma_{soil}$ as the maximum ordinate for the tributary area method.

PROBLEM 8.10 Pile Depth Calculation

The below 12 in diameter concrete pile has been placed into a soil with skin friction of 150 psf and a potential bearing capacity of 200 psf at the expected depth.

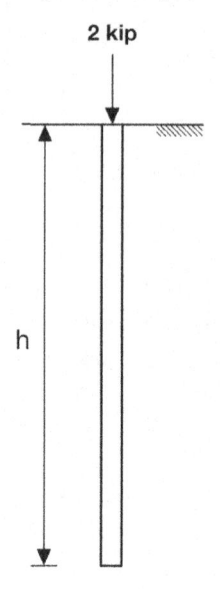

Pile depth h that should resist a pile loading of 2 kip with a safety factor of 2.5 is most nearly:

(A) 5 ft

(B) 10 ft

(C) 15 ft

(D) 20 ft

PROBLEM 8.11 Pile Load Carrying Capacity

A wooden pile is being driven into the earth using a drop hammer that weighs 5 kip with a free fall distance of 3 ft. The penetration that is measured per blow is 0.2 in.

The allowable carrying capacity for this pile is most nearly:

(A) 10,000 lb

(B) 15,000 lb

(C) 20,000 lb

(D) 25,000 lb

PROBLEM 8.12 *Pile Cap Design*

The below pile cap has five piles as shown with a vertical load applied to it $P = 150 \, kip$ and a moment in the x-y direction as shown of $M_{xy} = 75 \, kip \cdot ft$.

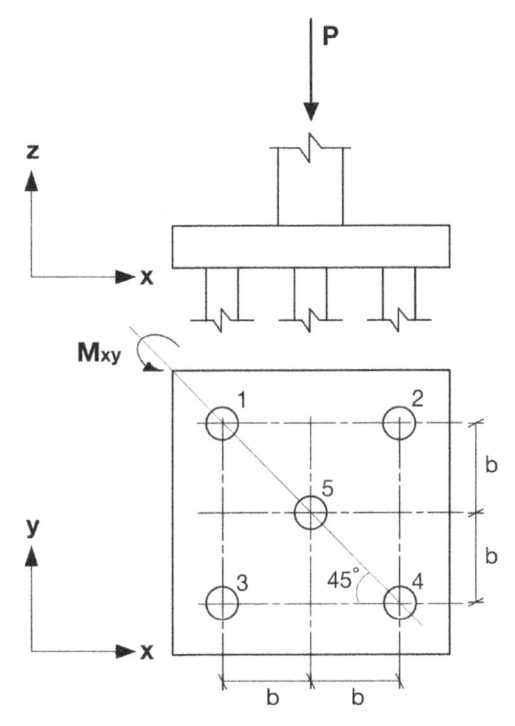

Given a single pile capacity of $Q = 40 \, kip$, the minimum dimension b between piles is most nearly:

(A) 35 in

(B) 70 in

(C) 105 in

(D) 140 in

PROBLEM 8.13 *Dynamic Pile Driving*

The following statement(s) describe the process of what happens when the hammer rams into the drivehead of a pile during a pile driving process:

I. The hammer transfers the force to the head of the pile over an infinite period of time.

II. A pulse is created, which travels down the pile in a wave shape.

III. The amplitude of the created wave decreases as it travels downwards and reaches the tip with a lesser amplitude.

IV. The force in the wave will reach the tip of the pile and pull the pile into the soil permanently.

V. The downward wave is reflected and backed up in the pile trying to push it upwards.

(A) I

(B) I, II, III

(C) II, IV, V

(D) II, III, IV, V

PROBLEM 8.14 *Ground Improvement*

A historic building in a densely populated urban area has started showing signs of settlement and structural distress.

The building is adjacent to other structures, making it crucial to minimize vibrations during any remedial work. The site has limited overhead space, and soil investigation reveals a mix of unsuitable fill material and dense layers at deeper depths.

The goal is to stabilize the building without causing damage to neighboring structures.

Based on the above, the following would be the best method to stabilize/underpin the foundations for this building:

(A) Compaction grouting with the use of compaction grout columns

(B) Jet grouting

(C) Helical piles

(D) Drilled micro-piles/mini piles

PROBLEM 8.15 *MSE External Stability*

The below retaining wall is a 21 ft high geotextile reinforced backfilled with $\gamma = 120\ pcf$ and $\emptyset = 35°$ granular material.

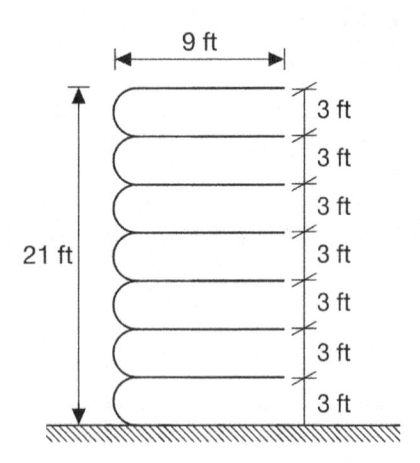

The safety factor for overturning for this wall is most nearly:

(A) 2.0

(B) 1.5

(C) 1.0

(D) 0.5

PROBLEM 8.16 *Slope Rapid Drawdown Factor of Safety*

The minimum Factor of Safety that should be used for rapid drawdown for new earth dams per the recommendations of the U.S. Army Corps of Engineers USACE 2003 is:

(A) 3.0

(B) 2.5

(C) 1.5

(D) 1.3

(✱) PROBLEM 8.17 *Retaining Tension Crack*

The below 20 ft high retaining wall supports soil with unit weight of $\gamma = 110\ pcf$, and a friction angle of $\emptyset = 35°$ along with a cohesion of $c' = 100\ psf$:

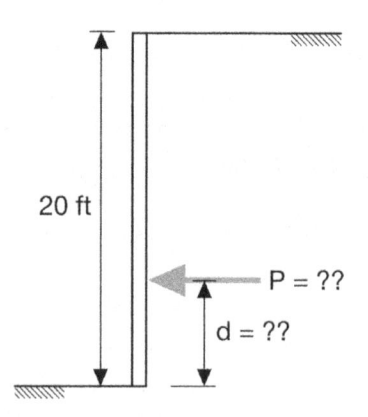

Assuming that pore water pressure for this soil is *zero*, the magnitude (P) and location (d) for the resultant lateral earth pressure behind the wall for a 1 ft strip are most nearly:

(A) $P = 4.0\ kip, d = 5.5\ ft$

(B) $P = 0.5\ kip, d = 3.5\ ft$

(C) $P = 0.1\ kip, d = 1\ ft$

(D) $P = 3.0\ kip, d = 16.5\ ft$

PROBLEM 8.18 *Retaining Wall Loading*

The below crane's front wheels and lever arms are located 5 ft away from the edge of a retaining wall. The front wheels and lever arms portion of load equals to 160 kip.

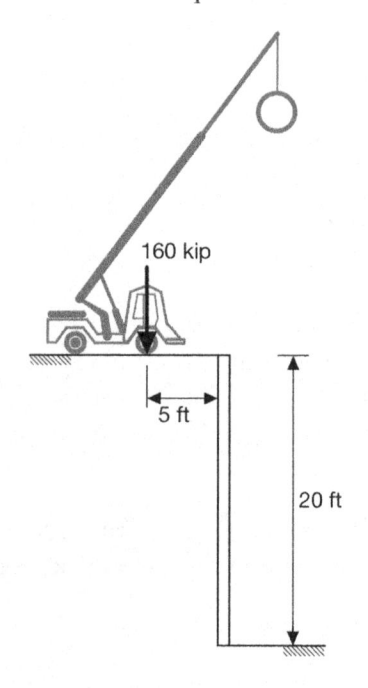

Assume the front wheels and lever arms act as a point load as shown. Based on this, the profile that represents the lateral pressure generated by the front wheels and lever arms alone is:

(A) Profile A

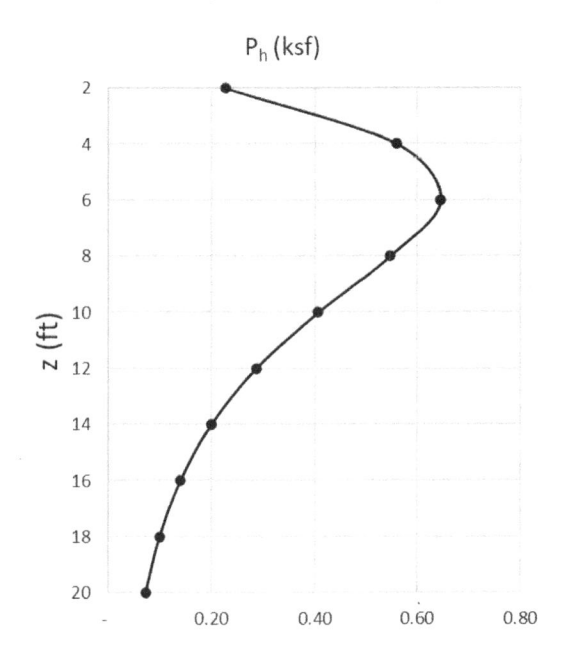

(B) Profile B

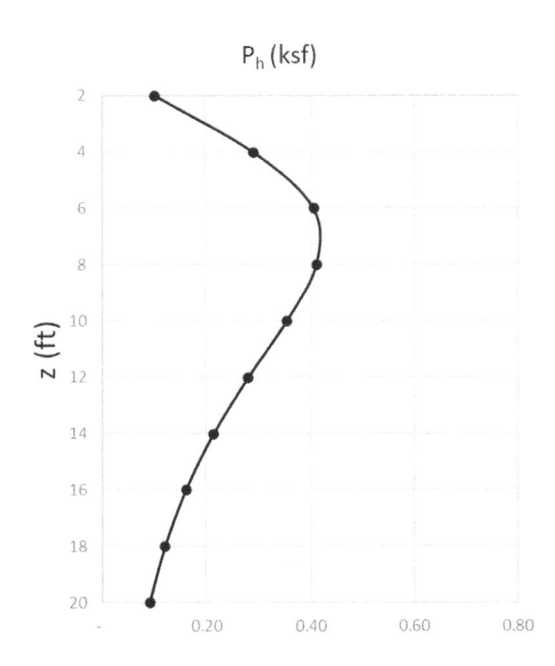

(C) Profile C

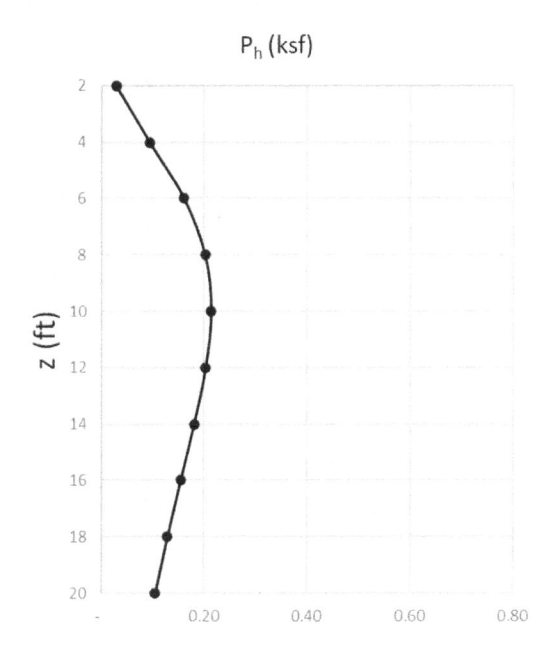

(D) Profile D

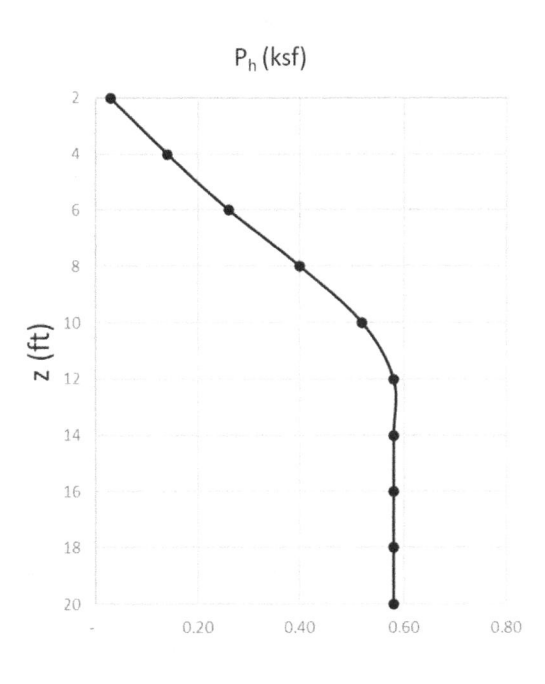

PROBLEM 8.19 Average Modulus of Elasticity

The below soil layers consist of difference moduli of elasticity bounded by a rock layer represented by the cross hatching.

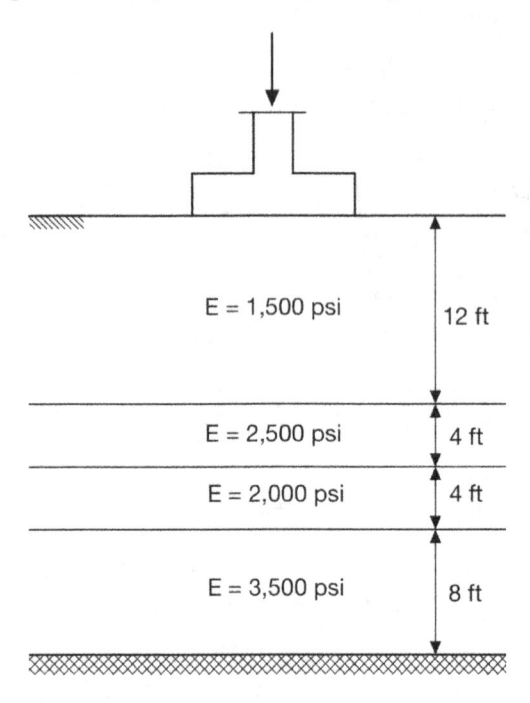

Layer	E	Thickness
1	$E = 1{,}500$ psi	12 ft
2	$E = 2{,}500$ psi	4 ft
3	$E = 2{,}000$ psi	4 ft
4	$E = 3{,}500$ psi	8 ft

For the purposes of calculating settlement, the average modulus of elasticity for the soil beneath the foundation is most nearly:

(A) 1,500 psi

(B) 2,015 psi

(C) 2,375 psi

(D) 2,285 psi

PROBLEM 8.20 Coulomb's Theory

The below gravity wall has the following wall and backfill properties:

- Backfill density $= 120$ pcf
- Gravity wall density $= 140$ pcf
- Backfill friction angle $= 35°$
- Wall/backfill friction angle $= 27°$

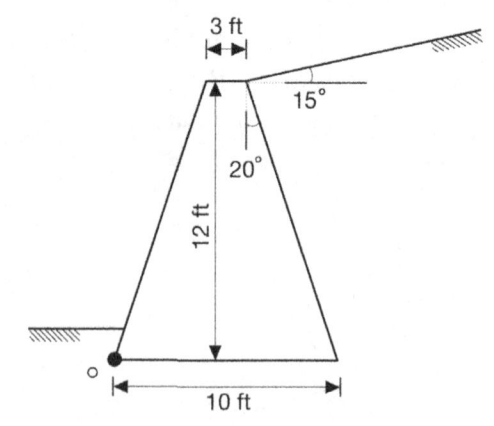

Based on the above information, and using Coulomb's theory to evaluate lateral earth pressures, the overturning moment in $kip \cdot ft$ per unit width of $1\ ft$ around point 'o' located at the bottom left corner of the wall is most nearly:

(A) 19.4

(B) 13.2

(C) 11.1

(D) 8.4

(✻) PROBLEM 8.21 Gravity Wall Stress

The below gravity wall has a density of $140\ pcf$, a backfill friction angle and density of $35°$ and $110\ pcf$ respectively, along with a wall/backfill friction angle of $21°$.

Using Coulomb Theory, the maximum stress showing in the below section is _____ psi.

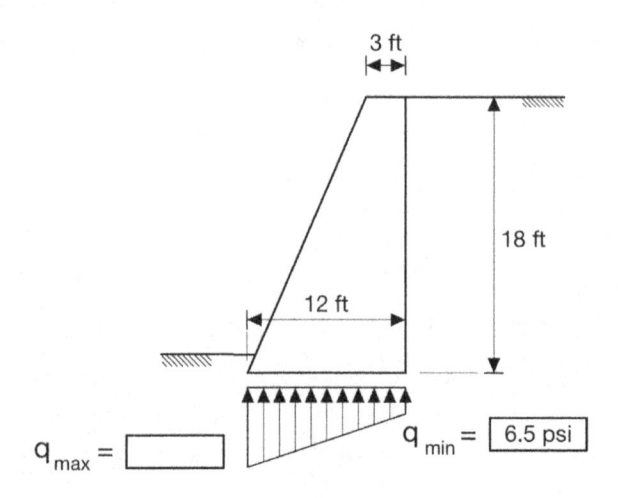

$q_{min} = 6.5$ psi

(⁂) **PROBLEM 8.22** *Rankine Various Layers*

The below soil cross section represents a retaining wall with two layers of soil backfill, each layer has different friction angle and soil density as follows:

- Layer 1:
  - Backfill density = 120 pcf
  - Backfill friction angle = 35°
- Layer 2:
  - Backfill density = 105 pcf
  - Backfill friction angle = 40°

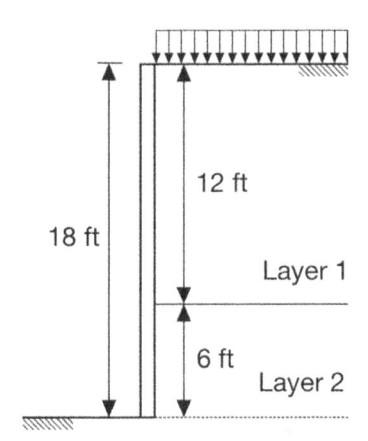

With a surcharge load of 200 psf as shown at the top side of the wall, and using Rankine theory, the overturning moment around the toe for this wall for a unit width of 1 ft is most nearly:

(A) 19 kip. ft per ft

(B) 27 kip. ft per ft

(C) 38 kip. ft per ft

(D) 48 kip. ft per ft

PROBLEM 8.23 *Anchored Sheet Pile*

The minimum distance d from the below retaining wall at which an anchor "deadman" can be placed in a way that permits the full development of passive pressure with a backfill friction angle of 37° and density of 115 pcf is most nearly:

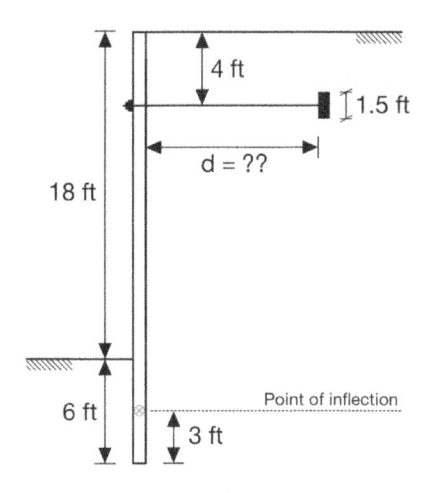

(A) 25.0 ft

(B) 20.0 ft

(C) 14.3 ft

(D) 10.0 ft

PROBLEM 8.24 *Trench Stress Evaluation*

The below trench is supported by two sheet piles and a hydraulic system. The supported soil consists of soft to medium clay with cohesion of 250 psf and density of 120 pcf.

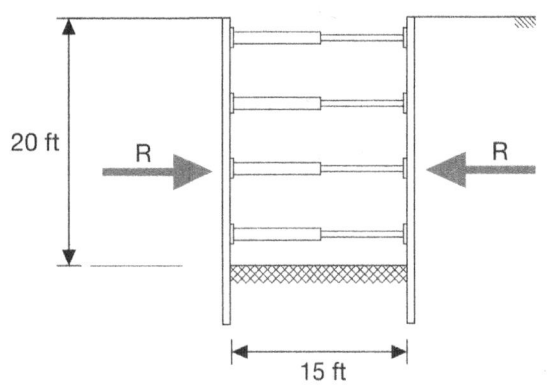

Given that the cut is supported by fixed ground as indicated by cross hatching, the lateral earth resultant forces R per unit width of 1 ft is most nearly:

(A) 24.5 kip per ft

(B) 18.0 kip per ft

(C) 14.0 kip per ft

(D) 10.5 kip per ft

PROBLEM 8.25 *Overburden Depth*

The minimum overburden depth that should be provided over the center of an anchor's grouted bond zone that is used to support a retaining wall is most nearly:

(A) 2.5 m

(B) 3.5 m

(C) 4.5 m

(D) 5.5 m

PROBLEM 8.26 *Sheet Pile Moment Diagram*

The below is a sheet pile embedded in the ground to support a trench with groundwater.

The showing pressure profile represents the resultant earth pressure acting on the sheet pile (*).

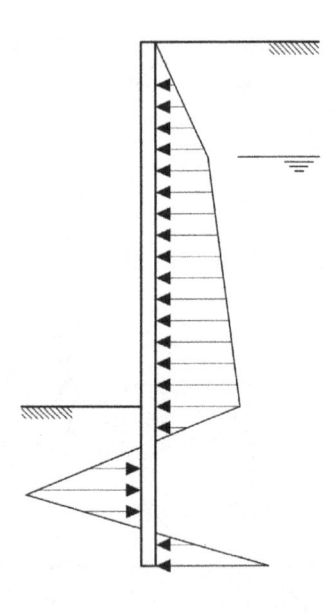

Based on the above information, the below profile is the best representation for the moment diagram that is generated from this earth pressure:

(A) Profile A

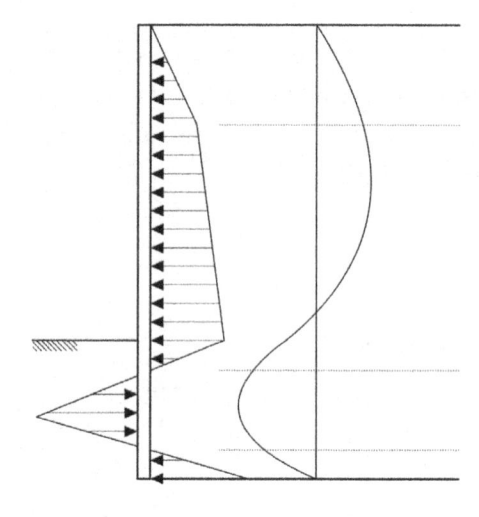

(B) Profile B

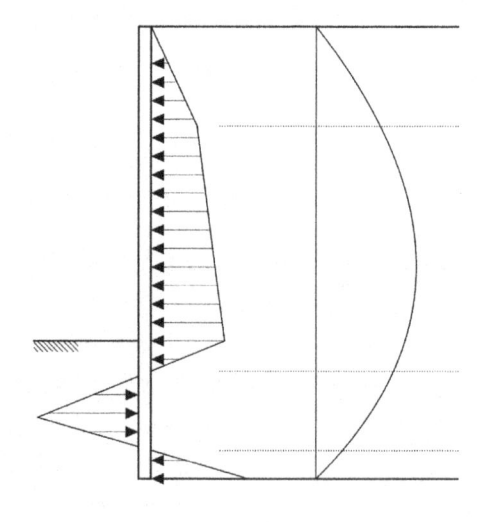

(C) Profile C

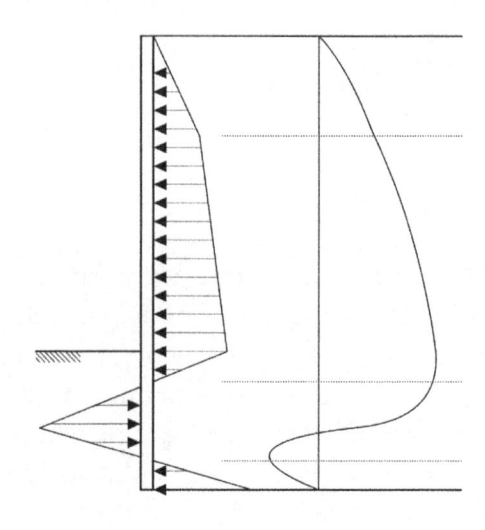

(D) Profile D

(D) Profile D

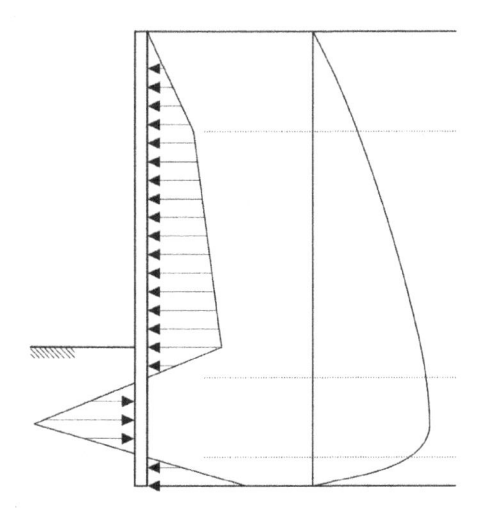

(*) An additional explanation at the end of this question's solution that explains how the resultant pressure profile is derived for sheet piles is provided for your own benefit.

PROBLEM 8.27 *Sheet Pile Cross Section*
The 15 ft high steel sheet pile, shown below, is used to support a deep trench. It is modeled as having a fixed end, with lateral earth pressure profile illustrated below:

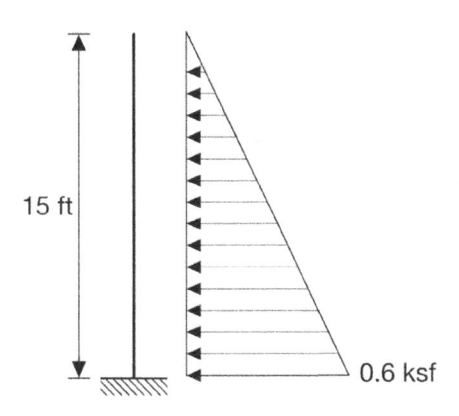

With a steel modulus of elasticity of $E = 29 \times 10^6$ psi, allowing a maximum deflection at the top free end of the sheet pile of 0.2 in, the following section is the most economical for use for this sheet pile arrangement (*):

(A) NZ 14
(B) NZ 21
(C) NZ 22
(D) NZ 26

(*) Refer to the *NCEES Handbook version 2.0*, Section 4.2.2 Steel Sheet Pile Properties for section properties.

PROBLEM 8.28 *Anchors Testing*
In a ground anchored wall system for a retaining wall, the number of anchors that should be tested after installation and prior to being placed into service is/are:

(A) Each anchor has to be tested.
(B) Every other anchor has to be tested.
(C) One every three anchors has to be tested.
(D) 50% of all anchors have to be randomly tested.

PROBLEM 8.29 *Pile Type Selection*
The most suitable type of pile for a location with numerous coarse gravel deposits is as follows:

(A) Precast concrete piles or drilled shafts
(B) H Piles
(C) Open end pipes
(D) Tapered piles or driven shafts

(⁂) **PROBLEM 8.30** *H Pile Depth*
The following steel H pile with a cross-sectional area of 53.2 in^2 is driven into cohesionless sand. The pile is driven in one sand layer that has no groundwater, a friction angle of $35°$ and density of 120 pcf.

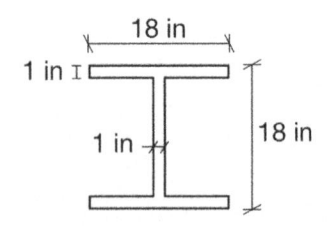

Ignoring pile toe resistance, the length/run of this H pile that can resist 25 tons by friction alone is most nearly:

(A) 15 ft

(B) 22 ft

(C) 27 ft

(D) 31 ft

PROBLEM 8.31 *Pile Capacity Change*
In the below earth cross section, the showing pile is 18 in diameter drilled concrete pile that crosses three layers of soil:

- Layer 1 (cohesionless soft soil)
  - Density = 80 pcf
  - Friction angle = 30°
- Layer 2 (cohesionless dense sand/granular soil)
  - Density = 105 pcf
  - Friction angle = 35°
- Layer 3 (cohesive clay)
  - Density = 120 pcf
  - Friction angle = 0
  - Cohesion = 500 psf

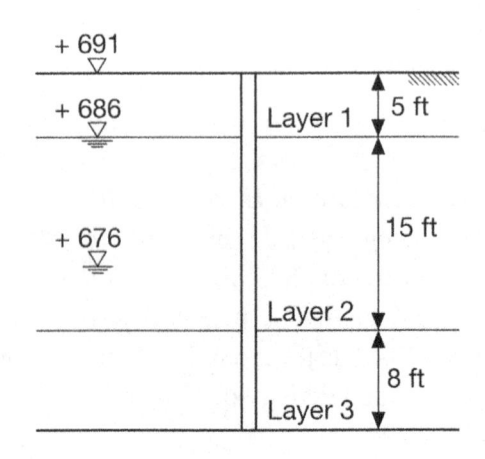

Due to groundwater depletion over time, the groundwater level dropped from +686 ft to +676 ft.

Using single pile capacity methods presented in NAVFAC DM-7.02, the change in this pile's ultimate compression capacity due to the drop in groundwater level is most nearly:

(A) 50 kip decrease

(B) 30 kip increase

(C) 20 kip decrease

(D) 10 kip increase

PROBLEM 8.32 *Pile Group*
Choose the only two correct statements from the below when it comes to assessing the capacity of a pile group:

☐ The capacity of a pile group is always less than the sum of the capacities of the individual piles forming the group.

☐ The capacity of a pile group is always equal to the sum of the capacities of the individual piles when installed in rock.

☐ In loose sand, the load-carrying capacity of an individual pile may increase due to the group effect.

☐ In granular soils, a pile group with spacing greater than seven times the average diameter, each pile can be considered as acting individually.

PROBLEM 8.33 *Pile Installation Methods*
Match the described site characteristic on the right with the required pile driving hammer on the left in the below table.

You may match one driving hammer type to more than one site characteristic.

Driving Hammer	Site or Soil Characteristic
Drop Hammer	Medium to hard ground
Air Hammer (double acting steam)	Small and inaccessible jobs
Diesel Hammer	Pile installed at 25-degree angle
Vibratory Hammer	Pile extraction
	Wet soils
	Granular soils or soft clays

(✻) PROBLEM 8.34 *Pile Lateral Deflection*

A reinforced concrete pile that is driven into a coarse-grained soil with a relative density of $Dr = 87\%$ and at a depth of $45\ ft$ has the following properties:

- Modulus of elasticity $E = 4 \times 10^6\ psi$
- Moment of inertia $I = 6,950\ in^4$

The pile is imbedded in a pile cap forming a rigid connection to the cap at the ground level.

The lateral force transmitted to a single pile from the pile cap is $P = 55\ kip$.

Given the above information, the estimated theoretical deflection that could occur at the ground level is most nearly:

(A) $1.2\ in$

(B) $0.7\ in$

(C) $0.05\ in$

(D) $0.3\ in$

PROBLEM 8.35 *Pile Static Testing*

In pile static testing, the offset limit line represents the following:

(A) A line on the load-displacement curve that measures allowable displacements during pile testing.

(B) A line representing the distance between the reference beam and the steel plate in the testing setup.

(C) A theoretical line outlining the sequence of testing for other piles during the testing.

(D) A line on the load-displacement curve below which the pile undergoes elastic deformation.

SOLUTION 8.1

Overturning:

Safety Factor against overturning is determined by calculating the Overturning Moment $\sum OM$ and the Resisting Moment $\sum RM$ around point 'O' as follows:

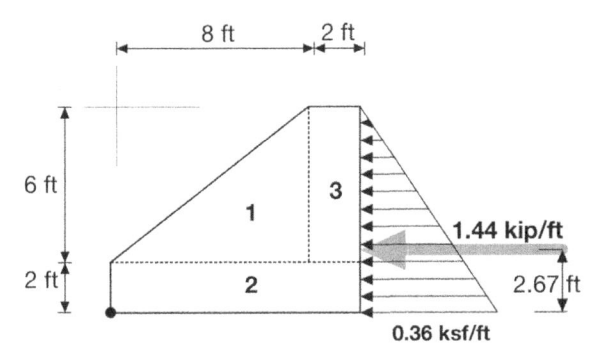

Resisting Moments $\sum RM$ as exhibited by the concrete wall:

Section	Volume	Wt.	Lever arm	R.M.
	ft^3/ft	kip/ft	ft	$kip.ft/ft$
1	24	3.6	5.33	19.2
2	20	3.0	5.0	15.0
3	12	1.8	9.0	16.2
Totals	56	8.4		50.4

$\sum RM = 50.4 \, kip.ft/ft$

$\sum OM = \left[\frac{1}{2} \times \left(45 \frac{psf}{ft} \times 8 \, ft\right) \times 8 \, ft\right] \times 2.67 \, ft$

$= 3,845 \, lb.ft/ft \, (3.8 \, kip.ft/ft)$

$FS_{OT} = \frac{\sum RM}{\sum OM} = \frac{50.4}{3.8} = 13.26$

Sliding:

Safety factor against sliding is determined by calculating the sliding force F_S and resisting force F_R as follows:

$F_S = 0.5 \times (45 \, psf/ft \times 8ft) \times 8ft \times \frac{1 \, kip}{1,000 \, lb}$

$= 1.44 \, kip/ft$

Resisting force is determined from the information given in the question regarding the cohesion exhibited by clayey silts which is $130 \, psf$. This value is multiplied by the bottom area of the wall. This value should not exceed half the deadload of the wall (*):

$F_R = 130 \, psf \times 1 \, ft \times 10 \, ft \times \frac{1 \, kip}{1,000 \, lb}$

$= 1.3 \, kip/ft$

$FS_{sliding} = \frac{F_R}{F_S} = \frac{1.3}{1.44} = 0.9$

Correct Answer is (A)

(*) Although this information is not required for the exam, it can be found in the IBC code Section 1806.3, Table 1806.2 and Section 1806.3.2.

SOLUTION 8.2

Locate the resultant force P_{total} by using the center of the first column as datum. Once found, double the distance to the center of the resultant $\bar{x}$ and add the remaining column 1 dimension twice to achieve the required length.

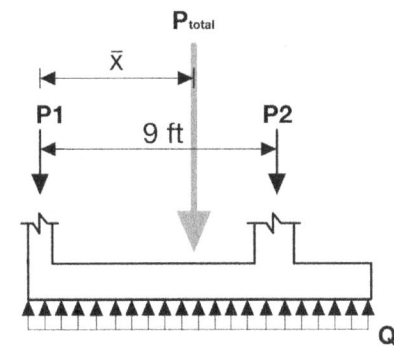

$\bar{x} \times P_{total} = 9 \, ft \times P2$

$\bar{x} = \frac{9 \, ft \times 75 \, kip}{75 \, kip + 50 \, kip} = 5.4 \, ft$

$L = 2 \times \frac{12 \, in}{2} \times \frac{1 \, ft}{12 \, in} + 2 \times (5.4 \, ft) = 11.8 \, ft$

$Q = \frac{P_{total}}{W \times L}$

$$W = \frac{P_{total}}{Q \times L} = \frac{75\ kip + 50\ kip}{2\ ksf \times 11.8\ ft} = 5.3\ ft$$

Correct Answer is (C)

SOLUTION 8.3
Service loads are used when checking for bearing pressures and area sizing. Ultimate loads are only used to design the concrete cross section.

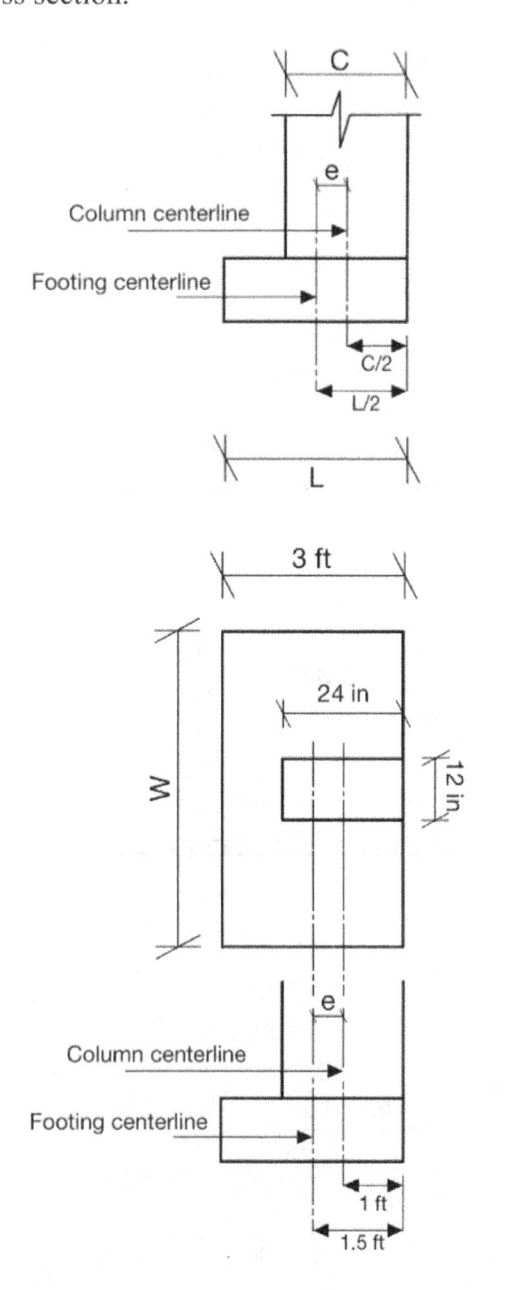

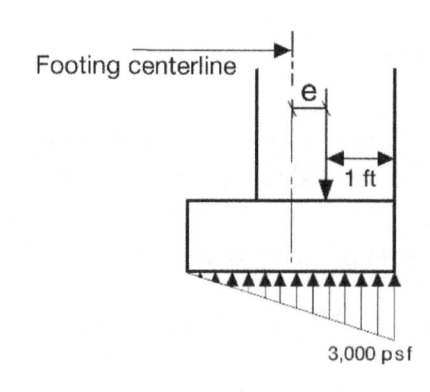

This is a case of eccentricity e, and the following equation can be used to ascertain that the lowest pressure in the eccentricity direction is zero, and the largest pressure does not exceed $3{,}000\ psf$.

$$Q = \frac{P}{A}\left(1 \mp \frac{6e}{L}\right)$$

First condition pressure $\geq$ zero:

$$\frac{P}{A}\left(1 - \frac{6e}{L}\right) \geq zero$$

$$1 - \frac{6e}{L} \geq zero$$

$$1 \geq \frac{6e}{L} \quad \rightarrow \quad e \leq L/6$$

$$e = \frac{L}{2} - \frac{C}{2}$$

$$\frac{L}{6} \geq \frac{L}{2} - \frac{C}{2}$$

$$\rightarrow L \geq \frac{3C}{2} = 36\ in\ (3ft)$$

$$\rightarrow e = 6\ in\ (0.5\ ft)$$

Second condition pressure $\leq 3{,}000\ psf$:

$$\frac{P}{A}\left(1 + \frac{6e}{L}\right) \leq 3{,}000\ psf$$

$$\frac{27\ kip}{3\ ft \times W}\left(1 + \frac{6 \times 0.5\ ft}{3\ ft}\right) \leq 3\ ksf$$

$$\rightarrow W \geq 6\ ft$$

Correct Answer is (B)

SOLUTION 8.4
The pressure right below the footing is calculated as follows:

$$q_o = \frac{P}{A} = \frac{65}{6 \times 6} = 1.8\, ksf\, (= 1,800\, psf)$$

Boussinesq's method:
Using the square footing part of Boussinesq's Isobars chart – copied below with permission from AASHTO for ease of reference, the horizontal and vertical coordinates of the chart are determined as portions of B (i.e., the footing width) as follows:

$$Horizontal\ axis = \frac{hor.\ location\ from\ edge\ of\ footing}{Width\ of\ the\ square\ footing}$$

$$= \frac{7.5\ ft}{6\ ft} = 1.25\ B$$

$$Vertical\ axis = \frac{Ver.location\ from\ bottom\ of\ footing}{Width\ of\ the\ square\ footing}$$

$$= \frac{12\ ft}{6\ ft} = 2\ B$$

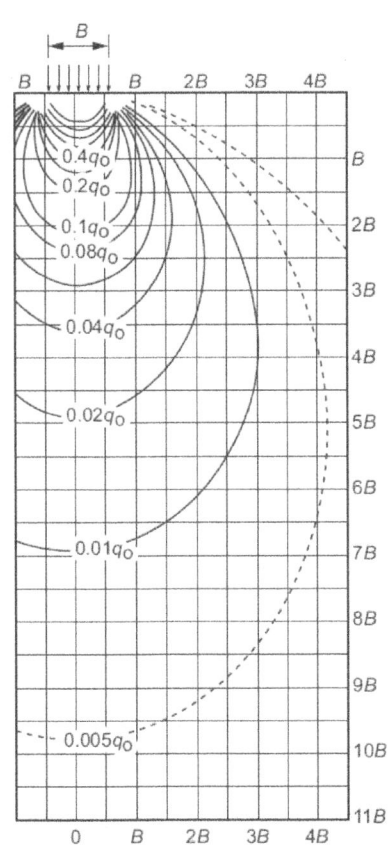

Interpolating those coordinates using the Isobar chart:

$$\Delta P = 0.06 q_o = 0.06 \times 1,800 = 108\ psf$$

The 2:1 method:
The 2:1 method assumes a 2:1 trapezoidal distribution of the load as shown in the following figure:

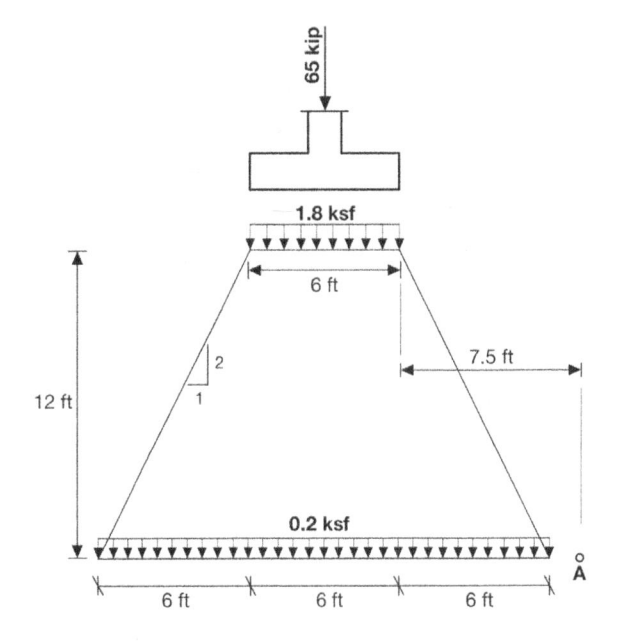

Based on the above distribution, a vertical depth of $Z = 12\ ft$ corresponds to a horizontal distance from the face of the footing of $6\ ft$. Point 'A' however sits at $7.5\ ft$ from the edge of the footing, i.e., $1.5\ ft$ away.

The pressure at point 'A' using this method is therefore *zero*.

Correct Answer is (A)

SOLUTION 8.5
Referring to the *NCEES Handbook version 2.0*, Section 3.5.2, the vertical elastic settlement is calculated as follows:

$$\delta_v = \frac{C_d \Delta p B_f (1-v^2)}{E_m}$$

C_d is the rigidity factor and is looked up from the table in the same section as '0.99' for rigid (i.e., concrete) square shaped foundations. Δp is the increase in pressure right below the foundation which is ($100 kip/36 ft^2$). B_f is the footing dimension which is $6 ft$ for squared footings in this case.

For fine medium dense sand, Poisson's ratio (v) is '0.25' as collected from the same section. The young modulus (E_m) shall be the lowest of the range provided in the guide – 120 tsf in this case – as the question is looking for maximum settlement.

$$\delta_v = \frac{0.99 \times \left(\frac{100}{36}\right) kip/ft^2 \times 6 ft \times (1-0.25^2)}{120 \frac{ton}{ft^2} \times 2 \frac{kip}{ton}}$$

$$= 0.064 \, ft \, (0.77 in)$$

Correct Answer is (A)

SOLUTION 8.6

The *NCEES Handbook,* Chapter 4 Geotechnical, Section 3.4.2.1 Bearing Capacity for Concentrically Loaded Square or Rectangular Footings, is referred to in order to provide a solution for this question.

First start with calculating the ultimate bearing capacity for the conditions provided in the question, then have it divided by the acting pressure from the column loading in order to determine the safety factor.

$$q_{ult} = c(N_c)s_c + q(N_q)s_q + 0.5\gamma(B_f)(N_\gamma)s_\gamma$$

The bearing capacity factors N_c, N_q and N_γ are collected from the table provided in the *NCEES Handbook* as 46.1, 33.3 and 48 respectively.

The shape correction factors s_c, s_q and s_γ are calculated using the following equations when $\emptyset > 0$:

$$s_c = 1 + \left(\frac{B_f}{L_f}\right)\left(\frac{N_q}{N_c}\right)$$

$$= 1 + \left(\frac{6}{6}\right)\left(\frac{33.33}{46.1}\right)$$

$$= 1.72$$

$$s_q = 1 + \left(\frac{B_f}{L_f} \tan\emptyset\right)$$

$$= 1 + \left(\frac{6}{6} \tan 35\right)$$

$$= 1.7$$

$$s_\gamma = 1 - 0.4\left(\frac{B_f}{L_f}\right)$$

$$= 1 - 0.4\left(\frac{6}{6}\right)$$

$$= 0.6$$

Given there is no surcharge load:
$$q = \gamma \, D_f$$

$$= 130 \, pcf \times 5 \, ft$$

$$= 650 \, psf$$

It is also important to remember that the density (γ) in the bearing capacity equation represents soil at the bottom of the footing, in which case the buoyant one will be used given that the bottom of the footing is submerged. It can either be calculated using the effective stress method as (γ') by deducting pore/water pressure from it, or a correction factor of '0.5' can be applied to it which can be collected from the Correction Factor table presented in the same chapter.

$$\gamma' = 130 \, pcf - 62.4 \, pcf$$

$$= 67.6 \, pcf$$

$$q_{ult} = c(N_c)s_c + q(N_q)s_q + 0.5\gamma'(B_f)(N_\gamma)s_\gamma$$

$$= 450(46.1) \times 1.72 + 650(33.3) \times 1.7$$

$$+ 0.5 \times 67.6(6)(48) \times 0.6$$

$$= 78{,}318.54\ psf$$

$$q_{actual} = \frac{650{,}000\ lb}{6\ ft \times 6\ ft}$$

$$= 18{,}055.56\ psf$$

$$FS = \frac{q_{ult}}{q_{actual}}$$

$$= \frac{78{,}318.54\ psf}{18{,}055.56\ psf}$$

$$= 4.34$$

Correct Answer is (A)

📄 SOLUTION 8.7

The *NCEES Handbook*, Chapter 4 Geotechnical, Section 3.4 Bearing Capacity can be referred in this question.

Bearing capacity equations of Section 3.4.2 for strip footings – copied below for ease of reference – which are fundamentally similar to bearing capacity equations for other footing types, has the total surcharge pressure at the base of the footing q as part of, and a major contributor to, the bearing capacity of the soil beneath. See below:

$$q_{ult} = c(N_c) + q(N_q) + 0.5\gamma(B_f)(N_\gamma)$$

$$q = q_{app} + \gamma_a D_f$$

Where (q_{app}) is the surcharge pressure at surface, which also has a positive impact on bearing capacities. (γ_a) is density of soil above the base of the footing, and (D_f) is the depth of the footing.

In conclusion, the removal of soil, or any other surcharge load on top of embedded foundations, by either excavation or scour, can substantially reduce the ultimate bearing capacity and may cause a catastrophic shear failure.

Correct Answer is (B)

(⭐) SOLUTION 8.8

Based on the information given in the question, Rankine's active coefficient is calculated as follows:

$$k_a = tan^2\left(45 - \frac{\emptyset'}{2}\right)$$

$$= tan^2\left(45 - \frac{37°}{2}\right)$$

$$= 0.25$$

The resultant soil pressure equation:

$$p_a = \frac{k_a\ h^2 \gamma_{soil}}{2} \quad \text{triangular shaped pressure}$$

$$p_a = k_a\ h_{above\ water}\ h_{below\ water} \gamma_{soil}$$
$$\text{rectangular shaped pressure}$$

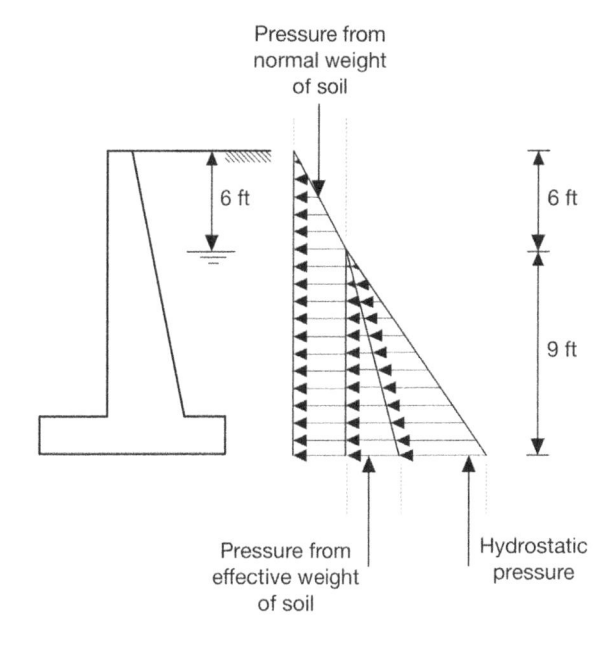

Based on this, lateral/overturning pressures are calculated per linear ft as follows:

Pressure resultant force from normal weight of soil:

$$p_{a,(0-6ft)} = \frac{1}{2} \times 0.25 \times (6ft)^2 \times 130\frac{lb}{ft^3}$$

$$= 585\ lb/ft$$

$$p_{a,(6-15ft)} = 0.25 \times 6\ ft \times 9\ ft \times 130\frac{lb}{ft^3}$$

$$= 1{,}755\ lb/ft$$

Pressure resultant force from effective weight of soil:

$$p_{a,(6-15ft)} = \frac{1}{2} \times 0.25 \times (9\,ft)^2 \times (130 - 62.4)\frac{lb}{ft^3}$$
$$= 684.5\,lb/ft$$

Pressure resultant force from hydrostatic pressures:

$$p_{water,(6-15ft)} = \frac{1}{2} \times (9ft)^2 \times 62.4\frac{lb}{ft^3}$$
$$= 2,527.2\,lb/ft$$

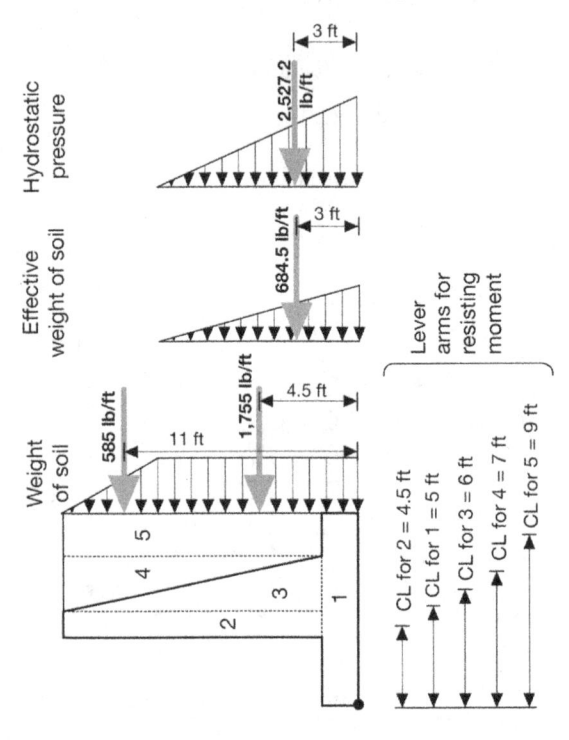

Overturning moments (O.M.) around the marked left most point:

Description	Force	Lever arm	Overturning Moment
	lb	ft	lb.ft
Soil 0-6 ft	585.0	11.0	6,435.0
Soil 6-15 ft	1,755	4.5	7,897.5
Eff. soil 6-15ft	684.5	3.0	2,053.5
Hydrostatic 6-15 ft	2,527.2	3.0	7,581.6
			23,967.5

The sum of overturning moments:

$$\sum O.M. = 23,967.5\,lb.ft/ft$$

Resisting moments (R.M.) with items 4 and 5 belong to the soil, $\gamma_{soil} = 130\,pcf$:

Sec	Area	Weight	Lever arm	Resisting Moment
	ft^2	lb	ft	$lb.ft$
1	17.5	2,625.0	5.0	13,125.0
2	13.25	1,987.5	4.5	8,943.8
3	19.9	2,985.0	6.0	17,910.0
4	19.9	2,587.0	7.0	18,109.0
5	26.5	3,445.0	9.0	31,005.0
				89,092.8

The sum of resisting moments

$$\sum R.M. = 89,092.8\,lb.ft/ft$$

Safety Factor calculation:

$$FS_{OT} = \frac{\sum R.M.}{\sum O.M.} = \frac{89,092.8}{23,967.5} = 3.7$$

Correct Answer is (C)

(✱) SOLUTION 8.9
The tributary area method is explained briefly in the NCEES Handbook Section 3.18.3 Anchor Loads.

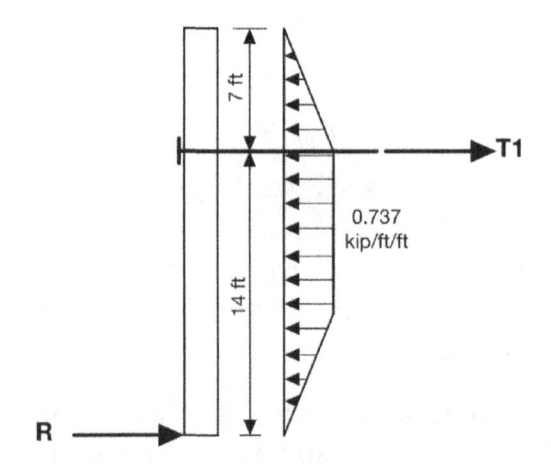

Using the traditional method explained in Solution 8.1 or 8.7 for lateral load calculations may give you indicative, not very accurate, results. The pressure behind retaining walls with anchors is complex and is best described per the below diagram (*).

$$k_a = tan^2\left(45 - \frac{\emptyset'}{2}\right)$$
$$= tan^2\left(45 - \frac{35°}{2}\right)$$
$$= 0.27$$

Maximum ordinate (p) – although given in the question – is available from the FHWA reference mentioned in the NCEES Handbook, and is calculated as follows:

$$p = k_a\, H\, \gamma_{soil}$$
$$= 0.27 \times 21\, ft \times 0.13\, kcf$$
$$= 0.737\, kip/ft/ft$$

Using the tributary area method explained in the NCEES Handbook Section 3.18.3 (*):

$$T1 = load\ over\ \left(H1 + \frac{H2}{2}\right)$$
$$= 0.5 \times 0.737 \times 7$$
$$\quad + 0.5 \times (14 \times 0.737)$$
$$= 7.74\, kip/ft$$

Given that anchors are placed at $3\,ft$ intervals, each anchor should have a capacity of **23.2 kip**.

A capacity of $23.2\,kip$ per anchor is considered significant. Another layer or two may be required in this case.

Correct Answer is (C)

(*) We assumed that the top and bottom triangular sections of the trapezoidal shaped pressure are $7\,ft$ high, which is a fair assumption for such a simple question.

The accurate method for estimating the pressure profile given in FHWA reference found in the NCEES Handbook is explained below briefly.

The trapezoidal loading per FHWA can be reconstructed in reference to the following diagram:

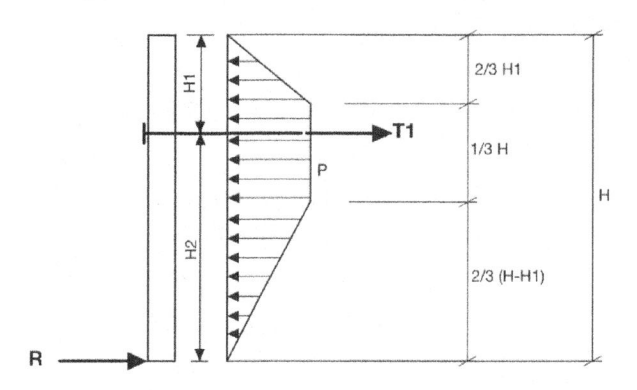

Based on the above, the following diagram is constructed to represent the pressure profile for the question in hand:

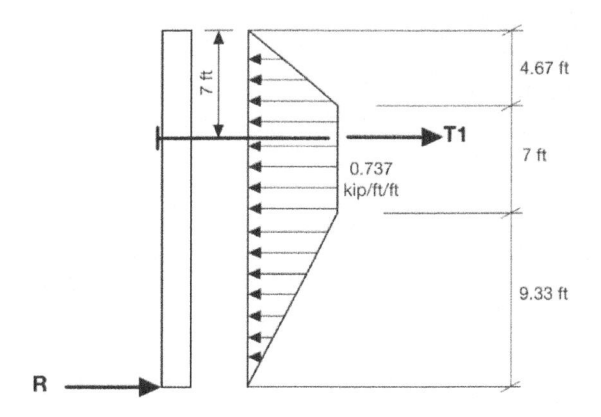

The tributary area contributing to the anchor loading in this case is:

$$T1 = load\ over\ \left(H1 + \frac{H2}{2}\right)$$

This area is better understood using the following diagram:

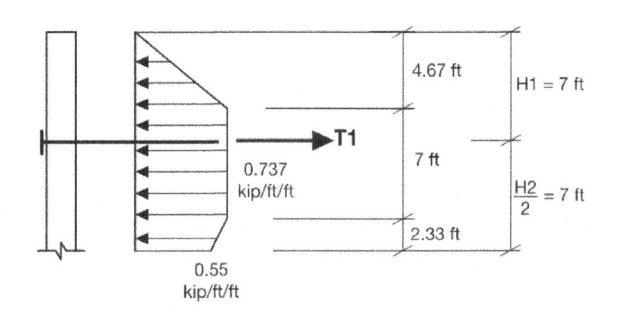

Load $T1$ is therefore calculated as follows:

$$T1 = \frac{0.737 \times 4.67}{2} + 0.737 \times 7 + \frac{(0.55 + 0.737)}{2} \times 2.33$$
$$= 8.37 \; kip/ft$$

Load per Anchor $= 8.37 \times 3 = 25.1 \; kip$

Reference to the above is as follows, which is reference A.[3] in the bibliography:

Federal Highway Administration, June 1999. Ground Anchors and Anchored Systems, Geotechnical Engineering Circular No.4. U.S. Department of Transportation.

SOLUTION 8.10
Determine: (1) skin friction with an unknown depth of h along with (2) the bearing capacity of soil, both shall equal to 2.5 times the required loading.

$Skin\; friction = 2\pi r \times h \times 150 \; psf$
$= 2\pi \times (½ \; ft) \times h \times 150 \; psf$
$= 150\pi h \; lb$

$Bearing\; resistance = \pi r^2 \times 200 \; psf$
$= \pi \times (½ \; ft)^2 \times 200 \; psf$
$= 50\pi \; lb$

$Total\; resistance = (150\pi h + 50\pi) \; lb$

$$S.F. = \frac{Resistance}{Actual\; Loading}$$

$$2.5 = \frac{150\pi h + 50\pi}{2000 \; lb}$$

$$h = 10.3 \; ft$$

Correct Answer is (B)

SOLUTION 8.11
The *NCEES Handbook version 2.0* Section 2.3.4.2 is used to solve this question using the modified Engineering-News Formula as follows:

$$P_{allowable} = \frac{2E_n}{S + K}$$
$$= \frac{2 \times (5{,}000 \; lb \times 3 \; ft)}{0.2 + 1.0}$$
$$= 25{,}000 \; lb$$

Correct Answer is (D)

SOLUTION 8.12
M_{xy} generates a moment couple between $Q_{2,M}$ and $Q_{3,M}$, $Q_{3,M}$ being the controlling pile with two compression forces acting on it.

$Q_{i,M}$ used in this question denotes the force generated at pile 'i' due to moment M_{xy}:

$$M_{xy} = \sqrt{2} \; b \; Q_{3,M} + \sqrt{2} \; b \; Q_{2,M}$$

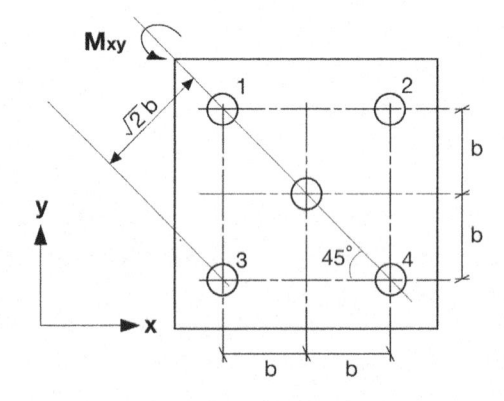

Since moment M_{xy} is applied at a $45°$ degree angle:

$$Q_{2,M} = Q_{3,M}$$

$$\rightarrow M_{xy} = 2\sqrt{2} \; b \; Q_{3,M}$$

$$Q_{3,M} = \frac{M_{xy}}{2\sqrt{2} \; b}$$

With five piles underneath the pile cap, the force generated at pile $Q_{3,P}$ is:

$$Q_{3,P} = \frac{150 \, kip}{5} = 30 \, kip$$

With a pile capacity of 40 kip:

$$Q_3 = Q_{3,P} + Q_{3,M}$$

$$40 \, kip = 30 \, kip + \frac{75 \, kip.ft}{2\sqrt{2} \, b}$$

$$b = 2.66 \, ft \, (32 \, in)$$

Correct Answer is (A)

SOLUTION 8.13

The *NCEES Handbook version 2.0* Section 2.3.4.1 and the figure presented in this section is used to understand the pile driving process. For more information on this, you can also refer to the *FHWA Soils and Foundations Reference Manual – Volume II, FHWA-NHI–06-089*, Section 9.9.5 Dynamic Analysis of Pile Driving.

The hammer transfers the force into the driving head of the pile over a finite period of time and the pulse created travels down in a wave shape, the amplitude of the wave decays due to the system damping before it reaches to the end of the pile. The force in the wave, however, will reach the tip/end of the pile and will pull the pile into the soil into a permanent set. The wave then is reflected upwards in the pile. After this reflection, an amount of permanent set or penetration of the pile tip remains in the earth.

The above makes only statements II, IV and V true and the rest is incorrect.

Correct Answer is (C)

SOLUTION 8.14

Drilled micro-piles is the best method for this case as they generate minimal vibrations compared to other methods. Moreover, they can be installed with limited overhead space and are effective in penetrating dense or obstruction-laden fills (*).

Correct Answer is (D)

(*) The below provides a brief explanation for various methods that can be used for underpinning. You are also encouraged to research and investigate the bibliography of this book for more methods:

Compaction Grouting (Compaction Grout Columns):
This method is used to reinforce unsuitable soils and support sinking structures or existing foundations. This technique can be performed with minimal overhead space (as little as 6 ft). The process involves inserting small-diameter steel casings to a specific depth and injecting a stiff grout at a high pressure as the casing is withdrawn. This displaces and compacts the surrounding soil, forming high-stiffness grout columns that, along with the improved soil, support the structure.

Jet Grouting:
This method is used to improve unsuitable soils to underpin foundations, construct earth support walls, and create groundwater cutoff walls. This technique involves advancing steel drill rods with a high-pressure water/grout jet. Once the desired depth is reached, a side-discharge grout jet (mostly rotating to cover $360°$) is activated, which erodes and mixes the soil to form columns with enhanced strength and reduced permeability.

Helical Piles:
Helical piles are deep foundation elements used for new foundations or underpinning existing ones. They generate no vibrations and can be installed in limited-access areas with only 6 ft of overhead space. These piles consist of galvanized steel shafts with screw-like bearing plates. Helical piles can function as end-bearing or side-friction elements.

Drilled Micro-piles (DMPs or Mini-Piles):
DMPs are high-capacity, small-diameter, drilled deep foundation elements used to support new foundations or underpin existing ones. They generate minimal vibrations and can be installed with only 8 ft of overhead space. DMPs typically consist of steel casing, threaded bar, and grout, deriving their capacity through side friction between the grout and surrounding soil or bedrock. Its use is more suitable when adjacent to vibration-sensitive structures while penetrating dense or obstruction-laden fill.

Chemical grouting:
This is a ground improvement technique used to stabilize granular soils and control water flow. This method involves injecting a low viscosity, non-particulate grout into the soil to fill voids and bind soil particles together, transforming the soil into a sandstone-like mass.

The process typically involves drilling small holes into the ground and injecting the chemical grout under pressure, which permeates the soil, fills voids, and binds particles.

Compared to compaction grouting, both methods aim to improve soil stability, chemical grouting however is more focused on permeating and binding soil particles, making it effective for water control and stabilization in tight spaces. Compaction grouting, on the other hand, is better suited for displacing and compacting soil to support structures and mitigate settlement

Fracture grouting:
Also known as compensation grouting. It involves injecting a cement slurry grout into the soil to create and fill fractures. It helps to lift and stabilize the overlying soil and structures.

Compared to compaction grouting, while both methods aim to stabilize soil, fracture grouting focuses on creating and filling fractures to stabilize the soil, making it effective for precise control and monitoring in tight spaces. Compaction grouting, on the other hand, is more suited for displacing and compacting soil to support structures and mitigate settlement.

You are encouraged to explore each of the above methods in more depth. Additionally, watching an online video per method can provide further insights into their applications.

SOLUTION 8.15

Safety Factor for overturning, or sliding, for MSE walls is determined in a similar fashion to any traditional retaining wall considering the wall depth is equivalent to the length of the embedded geotextile. See below sketch, and for context, you can follow the same logic presented in Solution 8.1 or 8.7 here.

Resisting Moment(s) $\sum RM$ is calculated as follows:

Section	Volume ft^3/ft	Wt. kip/ft	Lever arm ft	R.M. $kip.ft/ft$
Earth	189	22.7	4.5	102.2

$$\sum RM = 102.2 \; kip.ft/ft$$

For the overturning Moment(s) $\sum OM$, the lateral pressure is calculated as follows:

$$K_a = tan^2\left(45 - \frac{\emptyset}{2}\right)$$
$$= tan^2\left(45 - \frac{35}{2}\right)$$
$$= 0.27$$

$$\sigma_h = K_a \gamma H$$
$$= 0.27 \times 0.12 \; kcf \times 21 \; ft$$
$$= 0.68 \; ksf/ft$$

$$\sum OM = 0.5 \times \left(0.68\frac{ksf}{ft} \times 21 \; ft\right) \times 7 \; ft$$
$$= 50.0 \; kip.ft/ft$$

$$FS = \frac{\sum RM}{\sum OM} = \frac{102.2}{50.0} \cong 2.0$$

Correct Answer is (A)

SOLUTION 8.16

Reference is made to *UFC-3-220-10 Soil Mechanics,* Section 7-8 Required Factor of Safety for Soil Slopes, and Table 7-2.

Per the above table, the required minimum Factor of Safety for rapid drawdowns per USACE 2003 is '1.1' to '1.3'. Additionally, Solution 5.8 can be referred to here to understand the concept of rapid drawdowns and their effect on slopes.

Correct Answer is (D)

(✽) SOLUTION 8.17

Reference in this question is made to the *NCEES Handbook version 2.0,* Section 3.1.2 Rankine Earth Coefficient – Rankine Active and Passive Coefficients (Friction and Cohesion).

Normally retaining walls are backfilled with granular material. When backfill material has a certain amount of cohesion, the equation for lateral pressure becomes as follows:

$$P'_a = K_a(\gamma z - u) - 2c'\sqrt{K_a}$$

Rewriting this equation, taking into account that pore water pressure $u = 0$, and pressure $P'_a = 0$ at a critical depth location ($z_{critical}$) where the location of a tension crack would occur, provides the following:

$$0 = K_a(\gamma z_{critical}) - 2c'\sqrt{K_a}$$

$$\rightarrow z_{critical} = \frac{2c'\sqrt{K_a}}{\gamma K_a} = \frac{2c'}{\gamma\sqrt{K_a}}$$

$$K_a = tan^2\left(45 - \frac{\emptyset}{2}\right)$$
$$= tan^2\left(45 - \frac{35°}{2}\right)$$
$$= 0.27$$

$$z_{critical} = \frac{2 \times 100\ psf}{110\ pcf \times \sqrt{0.27}} = 3.5\ ft$$

The rest of the calculation, which relies on trigonometry, is presented on the following sketch along with the calculations for the resultant pressure of the bottom triangle which occurs upon forming the tension crack.

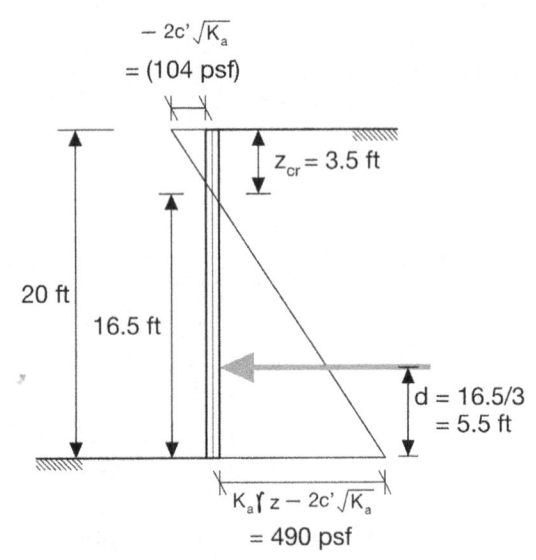

```
        -2c'√Kₐ
       = (104 psf)
          ├──┤
    ↑         ↑
    │        z_cr = 3.5 ft
    │         ↓
 20 ft
    │  16.5 ft
    │              ←────
    │              d = 16.5/3
    │                = 5.5 ft
    ↓    ├────────────┤
         Kₐγz − 2c'√Kₐ
         = 490 psf
```

The resultant force P equals to the area of the bottom triangle as shown on the sketch:

$$P = 0.5 \times (16.5\ ft \times 490\ psf/ft)$$
$$= 4{,}043\ lb/ft\ (4.0\ kip/ft)$$

Correct Answer is (A)

SOLUTION 8.18

Reference in this question is made to the *NCEES Handbook version 2.0*, Section 3.1.4 Load Distribution from Surcharge. The same information can also be found in *FHWA-NHI–06-089 Soils & Foundations Reference Manual – Volume II*, Section 10.4.3 Point, Line, and Strip Loads. The latter section contains a solved example as well.

While using the *NCEES Handbook version 2.0*, the figure relevant to the point load distribution is used along with the provided equations in page 78 next to the figure.

Using this information, and with a distance of $x = 5\ ft$ from the edge of the wall:

$$\bar{m} = x/H = 0.25 < 0.4$$

With this information, the first equation is used to determine the pressure as follows:

$$P_h\left(\frac{H^2}{Q_p}\right) = \frac{0.28\bar{n}^2}{(0.16 + \bar{n}^2)^3}$$

The above equation is already depicted in a figure for $\bar{m} \leq 0.4$ in the above indicated page of the handbook, also redrawn here for ease of reference:

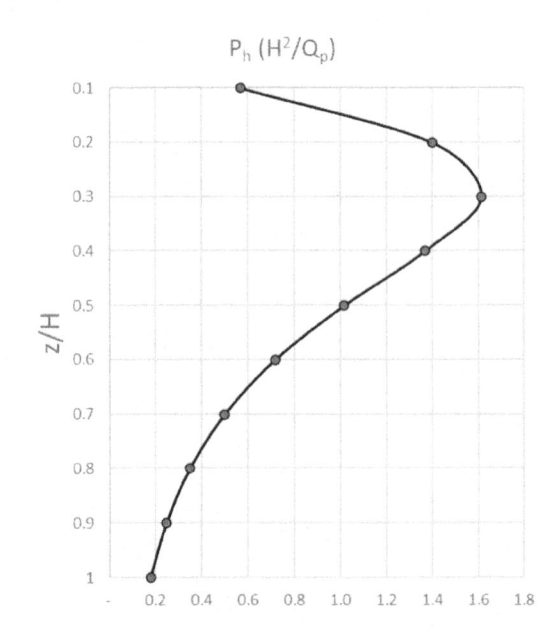

```
                    Pₕ (H²/Qₚ)
   0.1 ●
   0.2          ●
   0.3              ●
   0.4           ●
z/H 0.5       ●
   0.6      ●
   0.7    ●
   0.8   ●
   0.9  ●
   1  ●
     0.2 0.4 0.6 0.8 1.0 1.2 1.4 1.6 1.8
```

From this pressure profile, the maximum pressure occurs at z/H (or $\bar{n} = 0.3$) $\rightarrow z = 6\ ft$ taking datum from the top as indicated on these charts.

At this point, you can either use the above given equation to solve for (P_h), or use the x-axis ordinate from the given profile where it indicates that:

$$P_h\left(\frac{H^2}{Q_p}\right)_{@\bar{n}=0.3} \cong 1.6$$

$$\rightarrow P_{h\,@\,z=6\,ft} = \frac{1.6 \times 160\ kip}{(20\ ft)^2} = 0.64\ ksf$$

Based on this, **Profile A** would be the best match. The final pressure profile for profile A is computed below for convenience.

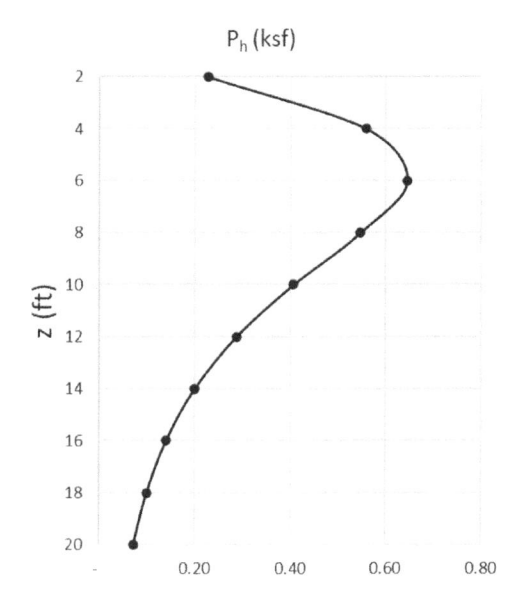

Correct Answer is (A)

SOLUTION 8.19

The best method for calculating the average modulus of elasticity is the weighted average method – i.e., taking the weight of each layer's thickness as follows:

$$E_{av} = \frac{1{,}500 \times 12 + 2{,}500 \times 4 + 2{,}000 \times 4 + 3{,}500 \times 8}{12+4+4+8}$$

$$= 2{,}285.7\ psi$$

Correct Answer is (D)

SOLUTION 8.20

Rankine theory for lateral earth pressures was used in many of the previous problems. Rankine theory assumes a smooth, vertical wall with horizontal backfills, which is suitable for straightforward applications.

This solution uses Coulomb theory as requested in the body of the question. Unlike Rankine's, Coulomb theory considers wall friction angle (δ), inclined backfills (β), and inclined walls as well (θ), making it more complex, providing a more comprehensive evaluation for lateral earth pressures.

See below sketch with the abovementioned angles defined:

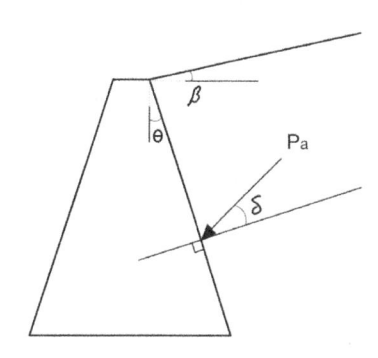

Coulomb's theory is presented in Section 3.1.3 Coulomb Earth Pressures of the *NCEES Handbook version 2.0*.

Based on the above, the below sketch represents the relationship between the earth pressure resultant force (P_a) and the wall for this question depicting the above discussed angles:

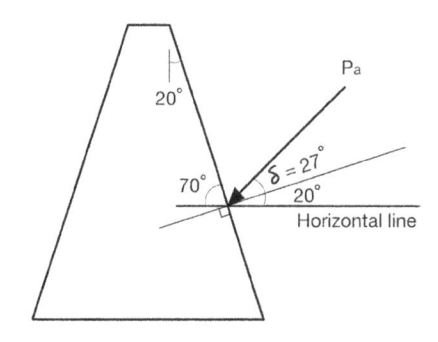

The active earth coefficient is calculated as follows:

$$k_a = \frac{\cos^2(\emptyset - \theta)}{\cos^2\theta \cos(\theta + \delta)\left[1 + \sqrt{\frac{\sin(\emptyset + \delta)\sin(\emptyset - \beta)}{\cos(\theta + \delta)\cos(\theta - \beta)}}\right]^2}$$

Where:

$\emptyset = 35°$ – Backfill friction angle

$\theta = 20°$ – Wall vertical angle

$\beta = 15°$ – Backfill angle

$\delta = 27°$ – Wall/Backfill friction angle

$$= \frac{\cos^2(35-20)}{\cos^2 20 \cos(20+27)\left[1+\sqrt{\frac{\sin(35+27)\sin(35-15)}{\cos(20+27)\cos(20-15)}}\right]^2}$$

$$= 0.56$$

Based on this, the resultant active pressure per unit width along with its horizontal component are calculated as follows:

$$P_a = 0.5 \times \gamma_{backfill} \times H^2 \times K_a$$

$$= 0.5 \times 120\ pcf \times (12\ ft)^2 \times 0.56$$

$$= 4{,}838.4\ lb\ (4.84\ kip)\ per\ ft$$

$$P_h = 4.84\ kip \times \cos(47)$$

$$= 3.3\ kip\ per\ ft$$

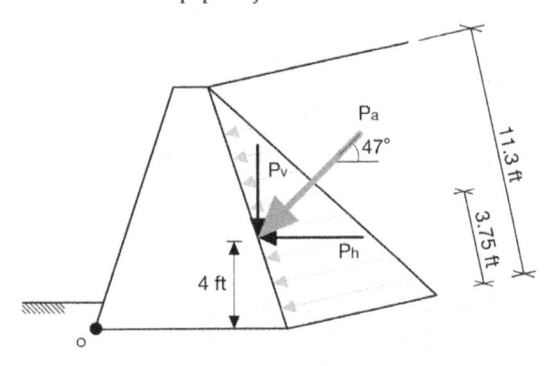

$$M = 3.3\ kip \times 4\ ft$$

$$= 13.2\ kip.ft\ per\ ft$$

Correct Answer is (B)

(✱) SOLUTION 8.21

The *NCEES Handbook version 2.0* page 76 is referred to, along with the methodology presented in Solution 8.20, to calculate lateral earth pressures.

The below cross section is used to find the eccentricity (*e*) for the stress below the wall, based upon which, the maximum and the minimum stresses can be calculated.

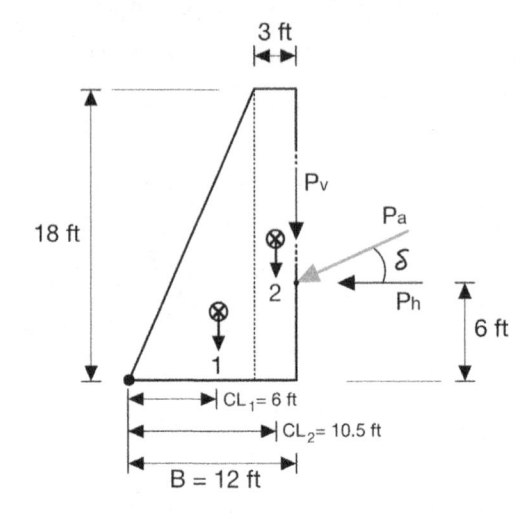

Start with calculating earth active pressures as follows:

$$k_a = \frac{\cos^2(\emptyset - \theta)}{\cos^2\theta \cos(\theta + \delta)\left[1+\sqrt{\frac{\sin(\emptyset + \delta)\sin(\emptyset - \beta)}{\cos(\theta + \delta)\cos(\theta - \beta)}}\right]^2}$$

Where:

$\emptyset = 35°$ – Backfill friction angle

$\theta = 0$ – Wall vertical angle

$\beta = 0$ – Backfill angle

$\delta = 21°$ – Wall/Backfill friction angle

$$= \frac{\cos^2(35-0)}{\cos^2 0 \cos(0+21)\left[1+\sqrt{\frac{\sin(35+21)\sin(35-0)}{\cos(0+21)\cos(0-0)}}\right]^2}$$

$$= 0.245$$

Based on this, the resultant active pressure per unit width along with its horizontal and vertical components are:

$$P_a = 0.5 \times \gamma_{backfill} \times H^2 \times K_a$$

$$= 0.5 \times 110\ pcf \times (18\ ft)^2 \times 0.245$$

$$= 4{,}365.9 \; lb \; (4.4 \; kip) \; per \; ft$$

$$P_h = 4.4 \; kip \times \cos(21)$$
$$= 4.1 \; kip \; per \; ft$$

$$P_v = 4.4 \; kip \times \sin(21)$$
$$= 1.6 \; kip \; per \; ft$$

Evaluate eccentricity by taking moment around the toe and equating the moments of the acting forces to the moments of the resultant forces (R) – see the preceding sketch. Please note that (d) represents the location of the resultant (R) from the toe – see the following sketch that depicts (R) as the resultant with (R_v) representing the resultant of vertical forces, and (R_h) as the resultant of horizontal/sliding forces:

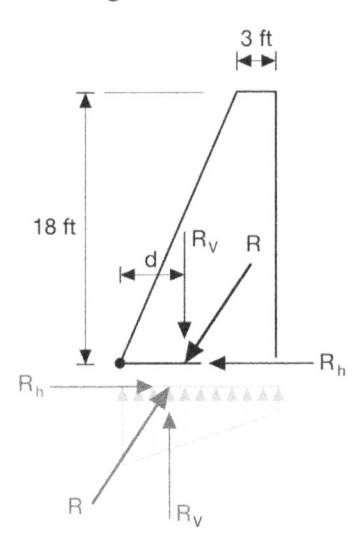

$$\sum M_{resultant} = \sum M_{acting \; forces}$$

$$R_v \times d = W_1 CL_1 + W_2 CL_2 + P_v B - \left(\frac{H}{3}\right) P_h$$

$$d = \frac{W_1 CL_1 + W_2 CL_2 + P_v B - \left(\frac{H}{3}\right) P_h}{R_v}$$

$$= \frac{W_1 CL_1 + W_2 CL_2 + P_v B - \left(\frac{H}{3}\right) P_h}{W_1 + W_2 + P_v}$$

Where:

$$W_1 = 0.5 \times 9 \times 18 \times 140$$
$$= 11{,}340 \; lb \; (11.34 \; kip) \; per \; ft$$

$$W_2 = 3 \times 18 \times 140$$
$$= 7{,}560 \; lb \; (7.56 \; kip) \; per \; ft$$

$$\rightarrow d = \frac{11.34 \times 6 + 7.56 \times 10.5 + 1.6 \times 12 - \left(\frac{18}{3}\right) \times 4.1}{11.34 + 7.56 + 1.6}$$

$$= 6.9 \; ft$$

Eccentricity is then calculated as follows (this equation can also be found in the referenced page of the NCEES handbook):

$$e = d - \frac{B}{2} = 0.9 \; ft$$

Based on the equations found in the above referenced material, q_{min} and q_{max} can be calculated as follows:

$$q_{min} = \frac{R_v}{B}\left(1 - \frac{6e}{B}\right)$$

$$= \frac{(11.34 + 7.56 + 1.6) \; per \; ft}{12 \; ft} \times \left(1 - \frac{6 \times 0.9 \; ft}{12 \; ft}\right)$$

$$= 0.94 \; ksf \; (6.5 \; psi)$$

$$q_{min} = \frac{R_v}{B}\left(1 + \frac{6e}{B}\right)$$

$$= \frac{(11.34 + 7.56 + 1.6) \; per \; ft}{12 \; ft} \times \left(1 + \frac{6 \times 0.9 \; ft}{12 \; ft}\right)$$

$$= 2.48 \; ksf \; (17.2 \; psi)$$

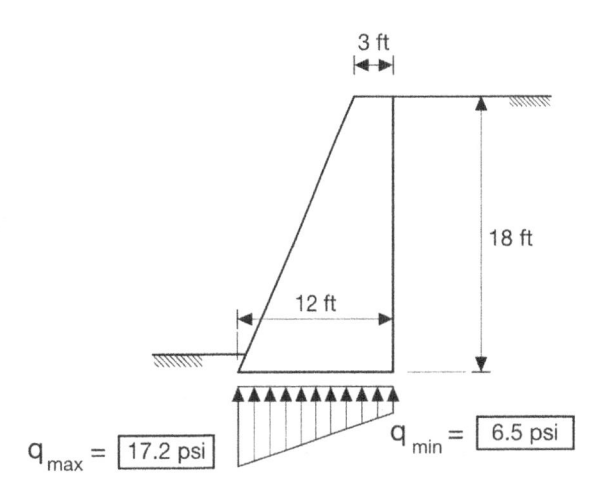

$q_{max} =$ 17.2 psi $q_{min} =$ 6.5 psi

(✻) SOLUTION 8.22

Start with calculating the Rankine coefficient for active pressure for the layers as follows:

$$K_{a1} = tan^2\left(45 - \frac{\emptyset}{2}\right)$$

$$= tan^2\left(45 - \frac{35}{2}\right)$$

$$= 0.27$$

$$K_{a2} = tan^2\left(45 - \frac{40}{2}\right)$$

$$= 0.22$$

The lateral stress is calculated at the top of each layer and at its bottom as follows:

Layer 1 with $K_{a1} = 0.27$ & $\gamma_1 = 120\ pcf$

At the top of the layer:

$$\sigma_{v1,z=0} = \gamma_1 z + surcharge$$

$$= 120 \times 0 + 200$$

$$= 200\ psf$$

$$p_{v1,z=0} = K_{a1} \times \sigma_{v1,z=0}$$

$$= 0.27 \times 200$$

$$= 54\ psf$$

At the bottom of the layer:

$$\sigma_{v1,z=12} = \gamma_1 z + surcharge$$

$$= 120 \times 12 + 200$$

$$= 1,640\ psf$$

$$p_{v1,z=12} = K_{a1} \times \sigma_{v1,z=12}$$

$$= 0.27 \times 1,640$$

$$= 442.8\ psf$$

Layer 2 with $K_{a2} = 0.22$ & $\gamma_2 = 105\ pcf$

At the top of the layer:

$$\sigma_{v2,z=0} = \gamma_2 z + \gamma_1 H_1 + surcharge$$

$$= 105 \times 0 + 120 \times 12 + 200$$

$$= 1,640\ psf$$

$$p_{v2,z=0} = K_{a2} \times \sigma_{v2,z=0}$$

$$= 0.22 \times 1,640$$

$$= 360.8\ psf$$

At the bottom of the layer:

$$\sigma_{v2,z=6} = \gamma_2 z + \gamma_1 H_1 + surcharge$$

$$= 105 \times 6 + 120 \times 12 + 200$$

$$= 2,270\ psf$$

$$p_{v2,z=6} = K_{a2} \times \sigma_{v2,z=6}$$

$$= 0.22 \times 2,270$$

$$= 499.4\ psf$$

Based on the above calculation, the following stress profile can be drawn showing the calculated resultant for each stress block:

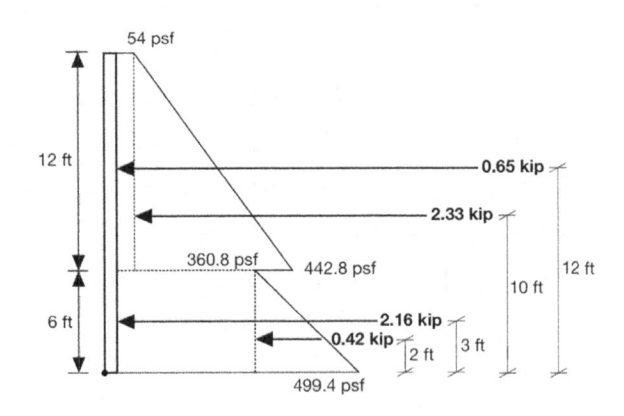

$$\sum M = 0.42\ kip \times 2\ ft + 2.16\ kip \times 3\ ft$$

$$+ 2.33\ kip \times 10\ ft$$

$$+ 0.65\ kip \times 12\ ft$$

$$= 38.42\ kip.ft\ per\ ft$$

Correct Answer is (C)

SOLUTION 8.23

Reference in this question is made to the *NAVFAC DM-7.02 Foundations and Earth Structures Design Manual*, Chapter 3, Figure 20 Design Criteria for Deadman Anchorage (*).

To achieve the full potential of an anchor in a way that develops the passive pressure for the sheet pile as requested in the question, the anchor should be placed behind the passive wedge as shown in the below sketch.

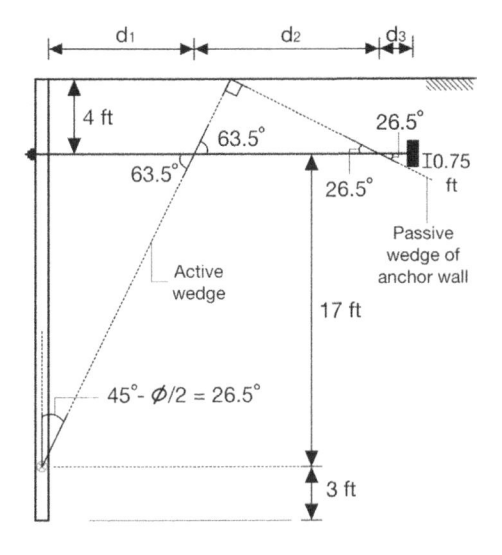

Based on the above, the minimum distance d is calculated as follows:

$$d = d_1 + d_2 + d_3$$

Using trigonometry:

$d_1 = 17/tan(63.5) = 8.5\ ft$

$d_2 = d_{2a} + d_{2b}$, see below:

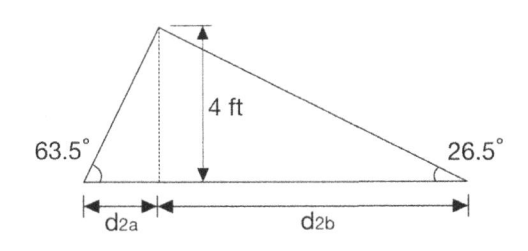

$d_2 = 4/tan(63.5) + 4/tan(26.5)$

$\quad = 10\ ft$

$d_3 = 0.75/tan(26.5)$

$\quad = 1.5\ ft$

$d = 8.5 + 10 + 1.5$

$\quad = 20\ ft$

Correct Answer is (B)

(*) As an additional resource, you can refer to the bibliography for some important references. Although these won't be provided during the exam, they offer a wealth of knowledge that you can benefit from. For this question, you can refer to the U.S. Army Corps of Engineers' EM 1110-2-2504 Design of Sheet Pile Walls added to the bibliography item A.[2].

SOLUTION 8.24

Reference in this question is made to the *NAVFAC DM-7.02 Foundations and Earth Structures Design Manual*, Chapter 3, Figure 26 Pressure Distribution for Brace Loads in Internally Braced Flexible Walls.

The second case is referred to here, in which case the sheet piles are applied to soft to medium clay.

To verify this case N_o should be > 6.0:

$$N_o = \frac{\gamma H}{c}$$

$$= \frac{120\ pcf \times 20\ ft}{250\ psf}$$

$$= 9.33 > 6.0 \rightarrow OK$$

Based on this confirmation, the following stress profile can be plotted as follows:

$$K_A = 1 - m\frac{4c}{\gamma H}$$

$$= 1 - 1 \times \frac{4 \times 250\ psf}{120\ pcf \times 20\ ft}$$

$$= 0.58$$

Where m is taken as '1' for stiff ground or '0.4 F_{SB}' for soft normally consolidate clay (which is not the case here). Where F_{SB} is factor of safety for bottom instability.

$$\sigma_h = K_A \gamma H$$

$$= 0.58 \times 120\ pcf \times 20\ ft$$

$$= 1{,}392\ psf\ (1.4\ ksf)$$

$$R = 15\ ft \times 1.4\ ksf + 0.5 \times 5\ ft \times 1.4\ ksf$$

$$= 24.5\ kip\ per\ ft$$

Correct Answer is (A)

SOLUTION 8.25

Reference in this question is made to the *NCEES Handbook version 2.0*, Section 3.18.4 Anchor Capacity, and page 212 (*).

According to the abovementioned reference, and to prevent grout leakage during the installation of pressure grouted anchors, a minimum overburden of **4.5 meters** above the center of the anchor bond zone is required. This also helps avoid ground surface heave due to high grouting pressures.

For gravity grouted anchors, this minimum overburden ensures sufficient soil pressure to develop the anchor's capacity.

Correct Answer is (C)

(*) You can also refer to *FHWA-IF-99-015 Geotechnical Engineering Circular No. 4: Ground Anchors and Anchored Systems*, Section 5.3.7 Spacing Requirements for Ground Anchors – reference A.[3] in the bibliography section. Although this reference will not be provided during the exam and is not listed in the required references, it is frequently mentioned and heavily referenced in the NCEES Handbook. It contains important information that you should familiarize yourself with prior to the exam.

SOLUTION 8.26

The following two methods detail the generation of the moment diagram for this sheet pile.

Additionally, at the end of this solution, you will find an explanation of how the pressure profile is constructed for sheet piles by understanding the regions of active and passive pressures around the pivot point (*).

The simple method:
The simplest way to determine the moment diagram is to consider the sheet pile as a fixed-end wall, with the portion below the ground anchoring it and creating a fixed-end moment condition. This implies that the most suitable diagrams are either profile C or D.

Since the moment does not change signs below the ground, particularly around the pivot point, profile C is excluded, making **Profile D** the best representation for the requested moment diagram.

The detailed method:
You can determine the moment diagram with the use of the shear diagram as shown in the following sketch:

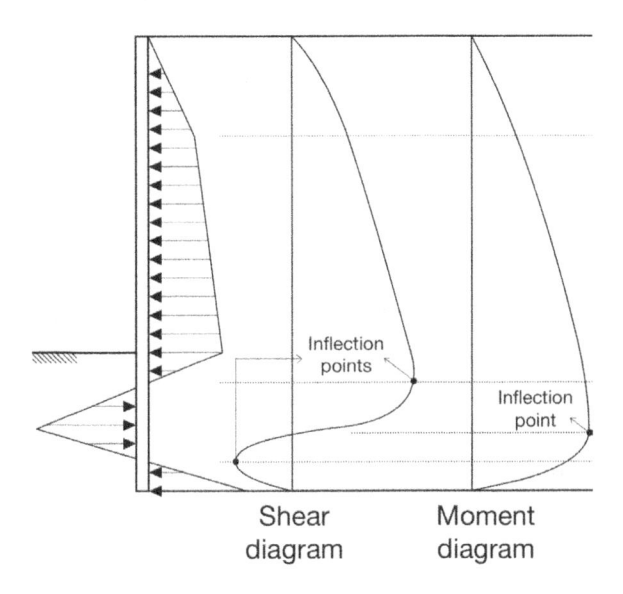

Determining the shear diagram first assists in:
- Identifying the location of maximum moments as they occur at zero shear locations.
- Determining whether there is a point of inflection in the moment diagram (i.e., changing directions), which occurs when shear changes signs.
- Establishing the shape function of the moment diagram. The moment diagram represents the area under the shear diagram, and its shape function is the integral of the shear diagram. Similarly, the shape of the shear diagram represents the integral of the load (or pressure profile) diagram.

The shear diagram:
Based on the above, it is important to observe that the pressure profile is a linear function. This indicates that the shape of the shear profile will be quadratic.

As the pressure profile changes signs (i.e., moving to the passive side of the sheet pile), this creates an inflection point in the shear profile, in other words, the shear profile changes direction.

In this example, the shear profile changes direction twice with two points of inflection and closes at the end of the sheet pile.

The Moment diagram:
As discussed, the shape of the moment diagram is the integral of the shear diagram. This means the moment diagram shape is cubic in this case, and the maximum moment will occur at the location where shear equals to zero. Additionally, the moment diagram will change direction as the shear diagram changes its sign creating an inflection point as discussed above, eventually, closing at the bottom of the sheet pile with zero moment.

The above discussion concludes that the best profile representing the moment diagram is **Profile D**.

Correct Answer is (D)

(*) Sheet Piles Pressure Profile:
The key feature of sheet piles, or cantilever walls, is that they use passive pressure to provide stability. It is important to understand that there is a pivot point; the wall cannot be in equilibrium if there is active pressure at the top and only passive pressure at the bottom, as this would cause the wall to rotate.

Equilibrium is achieved when there is a pivot point, allowing the wall to mobilize passive pressure in the opposite direction below the pivot point.

From this understanding, the below pressure diagram can be generated:

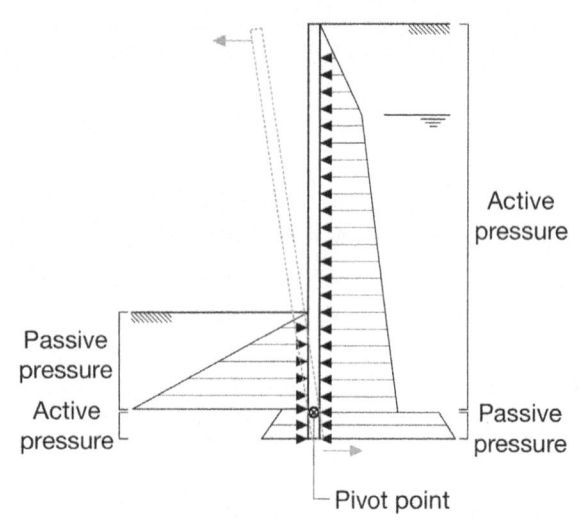

In the above diagram, the locations of passive and active pressures can be observed as follows:

o Above the pivot point and to the right side, the wall moves away from the soil, mobilizing active pressure.

o Below the dredge line and above the pivot to the left side, passive pressure is mobilized, which is much larger in value compared to the active pressure.

A similar logic applies to the soil below the pivot point where passive pressure is mobilized on the right side, which is larger than the active pressure generated to the left side of the wall.

Once these principles are established, the resultant pressure can be determined by subtracting the active pressure from the passive pressure on the opposite side, and vice versa generating a pressure profile that looks as follows:

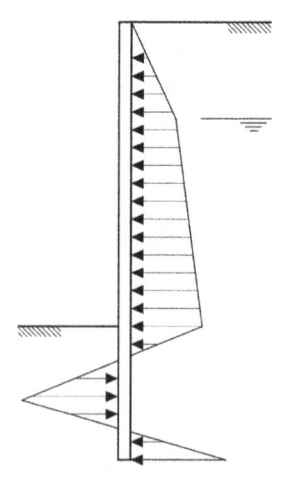

Determining the final pressure values is complex and requires several iterations to ensure that the sum of forces in the horizontal direction equals zero, and the moment summation at the bottom of the wall also equals zero. Once all is determined, maximum moment can then be computed for the purpose of designing the sheet pile section.

The USS Steel Sheet Piling Design Manual provides tables and diagrams to help resolve these calculations and determine the maximum moments needed to design the sheet pile system. This manual is not required in the exam (currently), it is only listed here and in the bibliography section for your information.

For more details on this, refer to bibliography item No. A.[7].

SOLUTION 8.27

The *NCEES Handbook version 2.0,* Shear, Moments and Deflection diagrams page 243 Case 18 is referred to as follows:

$$\Delta_{max} = \frac{Wl^3}{15EI}$$

Where W is the total load and is measured per unit width of $1\,ft$:

$$W = 0.5 \times \left(600 \tfrac{lb}{ft}/ft \times 15\ ft\right)$$
$$= 4{,}500\ lb\ per\ ft$$

$$0.2\ in = \frac{4{,}500\ lb/ft \times (15\ ft)^3 \times \left(\tfrac{12\ in}{ft}\right)^3}{15 \times 29 \times 10^6 \tfrac{lb}{in^2} \times I}$$

$$\rightarrow I = 301.7\ in^4\ per\ ft$$

Based on the four sections provided in the question, the section that has the closet moment of inertia, along with the least cross-sectional area, representing the lightest and most economical option, is highlighted in bold font in the table below:

Section	Cross-sectional Area	Moment of Inertia
	in^2/ft	in^4/ft
NZ 14	6.40	171.7
NZ 21	**7.80**	**313.4**
NZ 22	8.57	336.9
NZ 26	9.08	419.9

Correct Answer is (B)

SOLUTION 8.28
In anchored system applications, **each ground anchor undergoes testing after installation** and before being put into service with loads exceeding the design load. This testing process, along with specific acceptance criteria, ensures that the ground anchor can support the design load without significant deformations and that the expected load transfer mechanisms are properly established behind the critical failure surface. Once assessed, the ground anchor is stressed to a predetermined load and then locked off.

The above information can be found in *FHWA-IF-99-015 Circular No. 4: Ground Anchors and Anchored Systems*, Chapter 7 Load Testing and Transfer of Load to the Anchored System, (A.[3] in bibliography).

Although this reference is not listed in the required references, it is mentioned and heavily referenced in the NCEES Handbook. It contains important information that you should familiarize yourself with before the exam.

Correct Answer is (A)

SOLUTION 8.29
FHWA Soils and Foundations Reference Manual – Volume II, FHWA-NHI–06-089, Table 9.1 Pile type selection based on subsurface and hydraulic conditions.

When coarse gravel deposits are encountered, pile drivability can be challenging. **Therefore, using precast concrete piles or drilled shafts (option A) is the best choice**. In contrast, using H piles or open-ended pipes (options B and C) will result in longer, hence, costlier piles. Driven shafts or tapered driven piles will be difficult to install due to the high skin friction generated by the gravel deposits.

Correct Answer is (A)

(✱) SOLUTION 8.30
Reference is made in this solution to *FHWA Soils and Foundations Reference Manual – Volume II, FHWA-NHI–06-089*, Section 9.5.1 Ultimate Geotechnical Capacity of Single Piles in Cohesionless Soils/Nordlund Method. The same method is also presented in *FHWA Design and Construction of Driven Pile Foundations – Volume I, FHWA-NHI-16-009*, Section 7.2.1.3.1 Nordlund Method – Cohesionless Soils (*) (**).

Not considering pile toe resistance, with pile taper angle $\omega = 0$, Equation 9-6 of the first reference can be used as follows:

$$R_s = K_\delta C_F p_d \sin\delta C_d D$$

With pile depth D, and effective pressure midway p_d, this equation can be rewritten as follows:

$$R_s = K_\delta C_F \left(\frac{\gamma' D}{2}\right) \sin\delta C_d D$$

$$= 0.5 K_\delta C_F \gamma' D^2 \sin\delta C_d$$

$$\rightarrow D = \sqrt{\frac{2 R_s}{K_\delta C_F \gamma' \sin\delta C_d}}$$

Where:

$R_s = 25$ tons $(50{,}000$ lb$)$

$\frac{\delta}{\emptyset} = 0.93 \rightarrow \delta = 32.6°$

Using Figure 9-7 with $V = \frac{53.2\ in^2}{144\ in^2/ft^2} = 0.37\ ft^3/ft$

$K_\delta = 1.49$
Using Figure 9-10 or Table 9-6a.

$C_d = 18 \times 2 + 18 \times 2 = 72\ in\ (6\ ft)$ Check the comment under Step 6 of page 9-34 in reference to how to calculate perimeter (C_d) for H piles using the box method.

$C_F = 0.93$
Using Figure 9-12 for the correction factor C_F when $\delta \neq \emptyset$.

Putting it all together:

$$D = \sqrt{\frac{2 \times 50{,}000\ lb}{1.49 \times 0.93 \times 120\frac{lb}{ft^3} \times \sin(32.6°) \times 6\ ft}}$$

$= 13.6\ ft\ (*)$

Correct Answer is (A)

(*) Although we calculated the pile depth based on friction alone, below is the contribution from the toe resistance assuming the chosen pile depth is $D = 15\ ft$.

$$R_t = \alpha_t N'_q A_t p_t$$

Where:

$\alpha_t \approx 0.7$
Using Figure 9-13 with $D/b = \frac{15\ ft}{18\ in \times \frac{1\ ft}{12\ in}} = 10$ and $\emptyset = 35°$

$N'_q \approx 65$
Using Figure 9-13 with $\emptyset = 35°$

$p_t = 15\ ft \times 120\frac{lb}{ft^3} = 1{,}800\frac{lb}{ft^2}$
Which is the effective stress at toe.

Putting it all together:

$$R_t = 0.7 \times 65 \times 0.37\ ft^2 \times 1{,}800\frac{lb}{ft^2}$$

$= 30{,}303\ lb\ (15.2\ tons)$

This makes the total ultimate vertical capacity of this pile $\sim 40\ tons$.

(**) Attempting to solve the same problem using the β method. The effective stress β method is presented in section 9.5.2.2 of the same reference, and it can be used in both cohesionless and cohesive soils.

$$f_s = \beta p_o$$

Where f_s is shaft resistance or friction, p_o is the average effective stress $= \frac{D}{2}\gamma'$, and β can be collected from figure 9-17 with $\emptyset = 35°$ as 0.39 for sand.

Shaft resistance, is calculated as follows:

$$R_s = (perimeter \times D) \times \beta p_o$$

$$= perimeter \times \frac{D^2 \times \gamma'}{2} \times \beta$$

$$\rightarrow D = \sqrt{\frac{2R_s}{perimeter \times \gamma\prime \times \beta}}$$

$$= \sqrt{\frac{2 \times 50{,}000 \, lb}{8.67 \, ft \times 120\frac{lb}{ft^3} \times 0.39}}$$

$$= 15.7 \, ft$$

Where perimeter is taken here using all sides of the H pile section as $8.67 \, ft$ without using the box method recommended by Nordlund. The box method for computing H pile section perimeter is not mentioned or required in the β method.

The toe resistance using the β method can be calculated with the use of Equation 9-13 as follows:

$$R_t = N_t A_t p_t$$

Where:

$N_t \approx 55$ – using Figure 9-18 with $\emptyset = 35°$

$p_t = 15 \, ft \times 120 \frac{lb}{ft^3} = 1{,}800 \frac{lb}{ft^2}$
    Which is the effective stress at toe.

Putting it all together:

$R_t = 55 \times 0.37 \, ft^2 \times 1{,}800 \frac{lb}{ft^2}$

$= 36{,}630 \, lb \, (18.3 \, tons)$

SOLUTION 8.31

As requested in the question, reference is made to the *NAVFAC DM-7.02 Foundations and Earth Structures Design Manual*, Chapter 5 Deep Foundations, Figure 1 page 193 Load Carrying Capacity in Single Pile in Granular Soils, is used in this case.

Figure 2 of the same reference, which presents methods for cohesive soil layers (layer 3), are not used as the question requires the pile capacity change when groundwater drops from level 1 to 2 which occurs within layer 2 only.

Additionally, layer 1 with soft soil is ignored in all cases as recommended by the reference.

The ultimate load capacity in compression for single pile (without the tip resistance) is as follows:

$$Q_{ult} = \sum K_{HC} P_o \tan(\delta) S$$

Where:

K_{HC} for drilled piles less than $24 \, in$ is collected from the same figure as 0.7.

δ for concrete is collected from the same figure as $0.75 \times 35° = 26.25°$.

S is the surface area per unit length $= \left(\frac{18}{12} \times \pi\right) \times 1 \, ft = 4.71 \, ft^2 \, per \, ft$.

P_o is the average effective stress for the segment under consideration. This is calculated before and after groundwater depletion – Refer to Solution 2.2 to understand how to calculate effective stress if needed. Also, refer to Solution 8.22 to understand how to calculate stress at the top and bottom of layers.

Effective stress before groundwater depletion – layer 2 only:

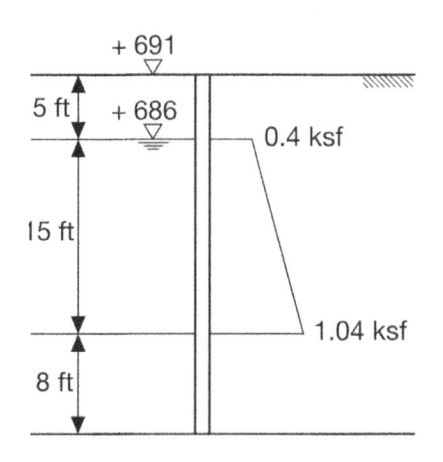

$$Q_{ult} = 0.7 \times \left(\frac{0.4+1.04}{2}\right) \times tan(26.25) \times (15 \times 4.71)$$
$$= 17.6 \; kip$$

Effective stress after groundwater depletion – layer 2 only:

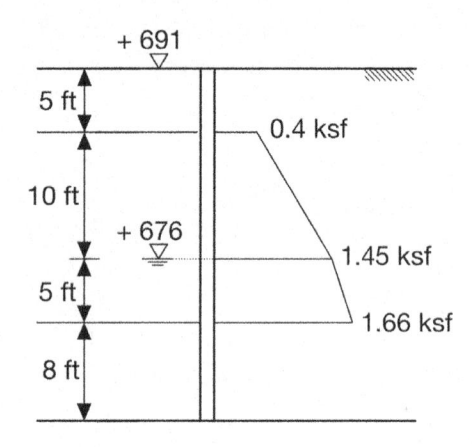

$$Q_{ult} = 0.7 \times \left(\frac{0.4+1.45}{2}\right) \times tan(26.25) \times (10 \times 4.71)$$
$$+ 0.7 \times \left(\frac{1.45+1.66}{2}\right) \times tan(26.25) \times (5 \times 4.71)$$
$$= 27.7 \; kip$$

As observed from the above, groundwater lowering causes pile compression capacity to increase by $27.7 - 17.6 = 10.1 \; kip \uparrow$.

Correct Answer is (D)

SOLUTION 8.32

Reference is made to the *NAVFAC DM-7.02 Foundations and Earth Structures Design Manual*, Chapter 5 Deep Foundations, Point 3, page 204, Bearing Capacity of Pile Group.

The bearing capacity of a pile group in soil is typically less than the sum of the capacities of individual piles within the group. This is termed as *group efficiency* which refers to the ratio of the capacity of a pile group to the combined capacities of individual piles at the same depth in the same soil deposit. This however does not apply for piles installed into rock formations.

The capacity of a pile group installed into rock is calculated by multiplying the number of piles by the individual capacity of each pile. However, if the formation is sloping, block failure must be considered, as sliding may occur along weak planes.

In cohesionless (granular) soils, **piles driven in a group act as individual piles if the spacing between piles is more than seven times the average pile diameter.** At closer spacings, they act as a group. The center-to-center spacing of adjacent piles should be at least twice the butt diameter.

In loose sand or gravel deposits, **the load-carrying capacity of an individual pile may increase when in a group due to densification during driving**.

Based on the above, the below are the only correct statements:

- ☑ In loose sand, the load-carrying capacity of an individual pile may increase due to the group effect.

- ☑ For a pile group with spacing greater than seven times the average diameter, each pile can be considered as acting individually.

SOLUTION 8.33

Reference is made to the *NAVFAC DM-7.02 Foundations and Earth Structures Design Manual*, Chapter 5 Deep Foundations, Section 4 Pile Installation and Load Tests.

The above chapter, specifically page 213, provides detailed information on installation methods based upon which, the required table can be reconstructed as follows:

Driving Hammer	Site or Soil Characteristic
Drop Hammer	Small and inaccessible jobs
Air Hammer (double acting steam)	Granular soils or soft clays
Diesel Hammer	Pile installed at 25-degree angle (Batter Piles)
Diesel Hammer	Medium to hard ground
Vibratory Hammer	Wet soils
Vibratory Hammer	Pile extraction

It is highly recommended that you scan the referenced chapter for more information on pile installation methods and the problems that can arise during, before or after piles' installations with the proposed solutions.

(✱) SOLUTION 8.34

Reference is made to the *NAVFAC DM-7.02 Foundations and Earth Structures Design Manual*, Chapter 5 Deep Foundations, Section 7 Lateral Load Capacity, Point 2 Deformation Analysis – Single Pile (*) (**) (***).

In reference to the above Chapter, along with Figure 10, this pile belongs to case II – Piles with Rigid Cap at Ground Surface.

Start with computing the relative stiffness factor T as follows:

$$T = \left(\frac{EI}{f}\right)^{1/5}$$

Where f is the coefficient of soil modulus variation and can be collected from Figure 9 for coarse grained soils with relative density $Dr = 87\%$ as $50\ tons/ft^3$.

Also, with paying attention to units in the above equation:

$$T = \left(\frac{4\times10^6\frac{lb}{in^2}\times 6{,}950\ in^4 \times\left[\frac{ft^2}{144\ in^2}\right]}{50\frac{tons}{ft^3}\times\left[\frac{2{,}000\ lb}{tons}\right]}\right)^{1/5}$$

$$= 4.5\ ft$$

Based on the given case, formulas from Figure 12 (the top figure) can be used to calculate deflection δ_p as follows:

$$\delta_p = F_\delta\left(\frac{PT^3}{EI}\right)$$

The deflection coefficient F_δ is collected from the x-axis of Figure 12 where the curve that should be used in this case belongs to $\frac{L}{T} = \frac{45}{4.5} = 10$ (the top curve). Since the deflection is required at depth $Z = 0 \rightarrow F_\delta \approx 0.92$.

$$\delta_p = 0.92 \times \left(\frac{55\ kip\times(4.5\ ft)^3}{4\times10^3\frac{kip}{in^2}\times 6{,}950\ in^4\times\left[\frac{ft^2}{144\ in^2}\right]}\right)$$

$$= 0.024\ ft\ (0.29\ in)$$

Correct Answer is (D)

(*) The solution provided here is based on the NAVFAC reference and the p-y curves method developed by Reese and Matlock. The reference to Reese and Matlock research is listed in Chapter 5 of NAVFAC as reference number 31 and in the bibliography here as reference number A.[6]. It is important to understand that p-y curves developed for one site do not apply to another. Which means, each site requires the development of their own p-y curves.

(**) Laterally loaded piles are very similar to beams on an elastic foundation. While a beam on an elastic foundation can be loaded at any point along its length, external loads on a pile are only applied at ground level.

Most numerical solutions for laterally loaded piles involve the concept of the modulus of subgrade reaction, which means that a soil medium may be approximated by a series of closely spaced independent springs. See below:

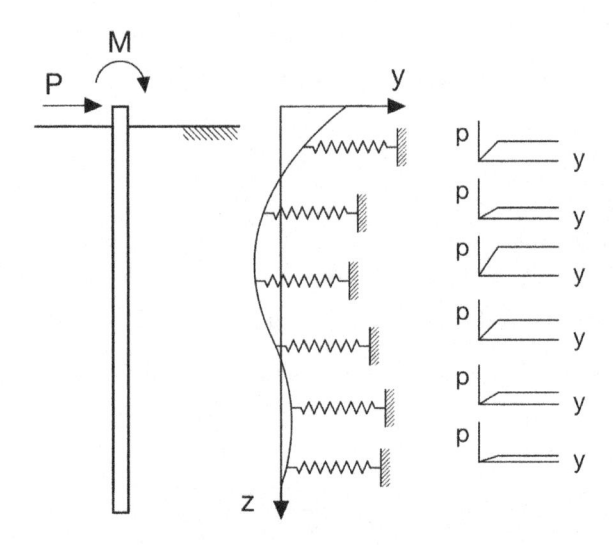

The key to solving lateral problems for piles rely on the determination of the modulus of subgrade reaction with respect to depth. Since the modulus of subgrade reaction can change with depth (except for overconsolidated soils as mentioned in NAVFAC), the coefficient of modulus variation (f) is introduced, which is collected from Figure 9 in NAVFAC of the same referenced chapter.

(***) The topic of lateral pile deflections is very broad. Because of this, make sure you familiarize yourself with the various lateral deformation methods, tests, and procedures presented in the following required NCEES references:

- FHWA Design and Construction of Driven Pile Foundations – Volume I, FHWA-NHI-16-009:
    - Section 7.3.7.3 p-y Method
    - Section 7.3.7.4 Strain Wedge Method
    - Section 7.3.7.5 Single Piles
    - Section 7.3.7.6 Pile Groups
- FHWA Design and Construction of Driven Pile Foundations – Volume II, FHWA-NHI-16-009:
    - Section 9.4 Lateral Load Test
- FHWA Drilled Shafts: Construction Procedures and Design Methods, FHWA-NHI-18-024:
    - Section 9.3 Design for Lateral Loading
- NAVFAC DM-7.02 Foundations and Earth Structures Design Manual:
    - Chapter 5 Deep Foundations Section 7 Lateral Load Capacity

SOLUTION 8.35

Reference is made to the *FHWA Design and Construction of Driven Pile Foundations – Volume II, FHWA-NHI-16-009*, Section 9.2.3 Presentation and Interpretation of Axial Compression Test Results, or Section 9.3.3 Presentation and Interpretation of Tension Test Results.

Based on the above, and in reference to Figure 9-8 for compression test (copied here for ease of reference and with permission from FHWA) or 9-11 for tension test of the above reference, it is observed that **the offset limit line is a line used to assess the performance of a pile under load**.

To further elaborate on this, during the test, a load is applied to the pile, and its response is recorded, as shown in the referenced figure below, in terms of displacement or settlement.

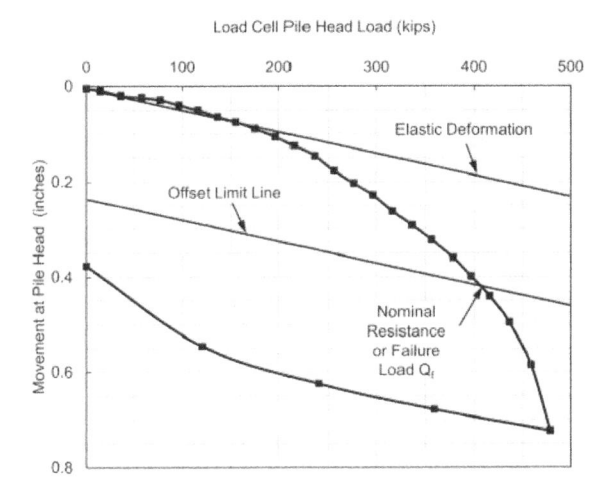

If the load-displacement curve value remains lower than this line, it indicates that the pile's displacement is within acceptable limits for the applied load. Conversely, if the curve values exceed this line, it suggests that the pile is experiencing excessive settlement, pull out, or failure under the given load, which may call for design adjustments.

Correct Answer is (A)

REFERENCES & BIBLIOGRAPHY

References

The following references have been used throughout the book and they are all required as indicated by the NCEES. You have to obtain and study them thoroughly prior to the exam:

1. **ASCE 7-16**
   Minimum Design Loads for Buildings and Other Structures, 2017, American Society of Civil Engineers, Reston, VA.

2. **EM 1110-2-1902**
   USACE Engineering and Design: Slope Stability, 2003, U.S. Army Corp of Engineers, Washington D.C.

3. **FHWA NHI-05-037**
   FHWA Geotechnical Aspects of Pavements, 2006, U.S. Department of Transportation, Federal Highway Administration, Washington, D.C.

4. **FHWA NHI-06-088**
   FHWA Soils and Foundations Reference Manual – Volume I, 2006, U.S. Department of Transportation, Federal Highway Administration, Washington, D.C.

5. **FHWA NHI-06-089**
   FHWA Soils and Foundations Reference Manual – Volume II, 2006, U.S. Department of Transportation, Federal Highway Administration, Washington, D.C.

6. **FHWA NHI-11-032 GEC No. 3**
   FHWA LRFD Seismic Analysis and Design of Transportation Geotechnical Features and Structural Foundations Reference Manual, 2011, Geotechnical Engineering Circulars, U.S. Department of Transportation, Federal Highway Administration, Washington, D.C.

7. **FHWA NHI-16-009 GEC No. 12**
   FHWA Design and Construction of Driven Pile Foundations – Volume I, 2016, Geotechnical Engineering Circulars, U.S. Department of Transportation, Federal Highway Administration, Washington, D.C.

8. **FHWA NHI-16-010 GEC No. 12**
   FHWA Design and Construction of Driven Pile Foundations. Volume II, 2016, Geotechnical Engineering Circulars, U.S. Department of Transportation, Federal Highway Administration, Washington, D.C.

9. **FHWA NHI-16-072 GEC No. 10**
   FHWA Geotechnical Site Characterization, 2017, Geotechnical Engineering Circulars, U.S. Department of Transportation, Federal Highway Administration, Washington, D.C.

10. FHWA NHI-18-024 GEC No. 10
FHWA Drilled Shafts: Construction Procedures and Design Methods, GEC No. 10 2018, Geotechnical Engineering Circulars, U.S. Department of Transportation, Federal Highway Administration, Washington, D.C.

11. NAVFAC DM-7.02
Foundations & Earth Structures, Design Manual 7.02, 1986, U.S. Army Corps of Engineers, Naval Facilities Engineering Command.

12. CFR TITLE 29 Part 1926
U.S. Department of Labor, Washington, D.C., July 2020. Part 1926 Safety and Health Regulations for Construction:
- Subpart CC, Cranes and Derricks in Construction,
- Part 1926:1400–1926:1442 with Appendix A–Appendix C
- Subpart E, Personal Protective and Life Saving Equipment, Part 1926.95–1926.107
- Subpart M, Fall Protection, 1926.500–1926.503 with Appendix A–Appendix E
- Subpart P, Excavations, 1926.650–1926.652 with Appendix A–Appendix F

13. UFC 3-220-05
Unified Facilities Criteria (UFC): Dewatering and Groundwater Control, 2004, U.S. Army Corps of Engineers, Naval Facilities Engineering Command. Air Force Civil Engineer Center, Washington D.C.

14. UFC 3-220-10
Unified Facilities Criteria (UFC): Soil Mechanics, 2022, U.S. Army Corps of Engineers, Naval Facilities Engineering Command, Air Force Civil Engineer Center, Washington D.C.

15. NCEES PE Civil Reference Handbook version 2.02

Bibliography

This section includes a broader list of important references either used in this book or that are important to check if you want to strengthen your knowledge in any of the areas.

This bibliography list is optional; however, it is highly advisable to check at least the ones referred to in the text of some of the question that you will encounter in this book.

A. Design of Earth Structures (Piles, Anchors, Sheet Piles, etc)

[1] Baligh & Abdelrahman (2006). Modification of Davisson's Method. Proceedings of the 16th International Conference on Soil Mechanics and Geotechnical Engineering.

[2] EM 1110-2-2504. Design of Sheet Pile Walls. U.S. Army Corps of Engineers, 1994.

[3] FHWA-IF-99-015. Ground Anchors And Anchored Systems. Geotechnical Engineering Circular No. 4.

[4] FHWA-IF-16-064. Design and Construction of Driven Pile Foundations – Comprehensive Design Examples. NHI courses No. 132021 and 132022.

[5] Reese et al (1984). Laterally Loaded Piles and Computer Program COM624G(U). Texas University at Austin.

[6] Reese & Matlock (1956). Non-Dimensional Solutions for Laterally Loaded Piles with Soil Modulus Assumed Proportional to Depth, Proceedings, Eighth Texas Conference on Soil Mechanics and Foundation Engineering, Austin, Texas.

[7] The USS Steel Sheet Piling Design Manual. United States Steel Updated and reprinted by U. S. Department of Transportation /FHWA with permission. July 1984.

B. Designing with Geosynthetics

[1] Koerner (2012). Designing with Geosynthetics 6th Edition Vol. 1 & 2. Library of Congress.

[2] Zornberg (2011). Advances in the Use of Geosynthetics for Waste Containment. Proceedings of the Fifteenth African Regional Conference on Soil Mechanics and Geotechnical Engineering, Maputo, Mozambique, July 17-21, pp. 1-21

[3] Zornberg & Christopher (2007). Chapter 37: Geosynthetics. In: The Handbook of Groundwater Engineering, 2nd Edition, Jacques W. Delleur (Editor-in-Chief), CRC Press, Taylor & Francis Group, Boca Raton, Florida.

C. Earthquake Engineering

[1] Kramer (1996). Geotechnical Earthquake Engineering. Prentice-Hall International (UK) Limited, London.

[2] Mononobe & Matsuo (1929). On The Determination Of Earth Pressures During Earthquakes. in Proceedings of the World Engineering Congress, p. 9, Tokyo, Japan, 1929.

[3] NCHRP Report 611. Seismic Analysis and Design of Retaining Walls, Buried Structures, Slopes, and Embankments. National Cooperative Highway Research Program.

D. General Material

[1] Das, B. (2008). Advanced Soil Mechanics Third Edition. Taylor & Francis.

[2] Day (2009). Foundation Engineering Handbook. Design and Construction with the 2009 International Building Code. ASCE Press.

[3] Das (2011). Principles of Foundation Engineering Seventh Edition. Cengage Learning. USA.

[4] Design Standard No. 13. Embankment Dams. Reclamation Managing Water in the West. Chapter 8 – Seepage. U.S. Department of Interior Bureau of Reclamation, Jan 2014.

[5] Foundation and Earth Structures Design Manual. NAVFAC DM 7.1, U.S. Army Corps of Engineers, Naval Facilities Engineering Command.

[6] Oyinkanola (2016). Correlation Between Soil Electrical Resistivity and Metal Corrosion Based on Soil Types For Structural Designs. Scientific Research Journal (Scirj), Volume Iv, Issue I.

[7] Terzaghi et al. (1996). Soil Mechanics in Engineering Practice. John Wiley & Sons, Inc.

[8] Okabe (1926). General Theory of Earth Pressures. Journal of the Japan Society of Civil Engineering, vol. 12, no. 1.

[9] Venkatramaiah (2006). Geotechnical Engineering. New Age International Publishers.

E. Ground Improvement

[1] Ayele (2017). A Case Study on Ground Improvement Techniques and its Applications. International Journal of Scientific & Engineering Research Volume 8, Issue 9.

[2]   Bui et al. (2023). A Case Study of the Improvement of an Old Building's Shallow Foundation Resting on Sandy Clay Soil Using Cement Slurry Grouting Technique. E3S Web of Conferences.

[3]   Evans (2022). Fundamentals of Ground Improvement Engineering. Deanta Global Publishing Services, Chennai, India.

[4]   Froehlich (2013). Protecting Bridge Piers with Loose Rock Riprap. Journal of Applied Water Engineering and Research. Vol. 1, No. 1, 39–57.

[5]   FHWA-NHI-16-027. Ground Modification Methods Reference Manual – Volume I. U.S. Department of Transportation Federal Highway Administration.

[6]   FHWA-NHI-16-028. Ground Modification Methods Reference Manual – Volume II. U.S. Department of Transportation Federal Highway Administration.

[7]   Tsitsas et al (2019). Use of Compaction Grouting as Ground Improvement Technique in Compressible Solid Waste Landfill, Proceedings of the XVII ECSMGE-2019, Geotechnical Engineering Foundation of the Future.

[8]   Winter (1998). Case Histories of Soil Improvement, Grouting, Geosynthetics, Dynamic Compaction, Vibroflflotation, Blasting and Other Methods Including Geo Economics. Fourth International Conference on Case History in Geotechnical Engineering St. Louis, Missouri.

[9]   UFC 3-220-06. Unified Facilities Criteria (UFC): Soil Mechanics, 2022, U.S. Army Corps of Engineers, Naval Facilities Engineering Command, Air Force Civil Engineer Center, Washington D.C.

[10] Yang et a. (2019). Application of Post-Grouting in Bridge Foundation Reinforcement: A Case Study. Journal of GeoEngineering, Vol. 14, No. 3, pp. 155-165.

This page is intentionally left blank

Your Feedback Matters – Make Sure You Share it With Others

Good day,

As you reach the final pages of this book, we would like to express our sincere gratitude for choosing it as your guide to aid you in your journey toward success in the PE exam. We have poured countless hours into meticulously crafting the questions and practice exams within these pages.

Your opinion matters greatly in helping others discover the value of this resource. If you found this book beneficial, kindly consider leaving your positive and honest feedback on the platform that you bought it from - like Amazon. Your words will not only acknowledge the hard work invested into producing this book but will also guide future readers in their quest for quality study materials.

Remember, your review is more than just feedback; it's a beacon for those seeking reliable resources. Your support can make a significant difference, ensuring that this book continues to assist aspiring professionals on their path to success.

Thank you for being a part of this journey, and we appreciate your commitment to sharing your experience with others.

PE ESSENTIAL GUIDES

Made in the USA
Coppell, TX
27 April 2025